53.00

PHOS 0592

54

S0-AFS-157

Radionavigation systems

Radionavigation systems

Börje Forssell

Prentice Hall
New York   London   Toronto   Sydney   Tokyo   Singapore

First published 1991 by
Prentice Hall International (UK) Ltd
66 Wood Lane End, Hemel Hempstead
Hertfordshire HP2 4RG
A division of
Simon & Schuster International Group

© Prentice Hall International (UK) Ltd, 1991

All rights reserved. No part of this publication may be reproduced,
stored in a retrieval system, or transmitted, in any form, or by any
means, electronic, mechanical, photocopying, recording or
otherwise, without prior permission, in writing, from the publisher.
For permission within the United States of America contact Prentice
Hall Inc., Englewood Cliffs, NJ 07632.

Typeset in 10/12 pt Times
by KEYTEC, Bridport, Dorset

**Printed in the United States of America
by Halliday Lithograph.**

Library of Congress Cataloging-in-Publication Data

Forssell, Börje, 1939–
    Radionavigation systems/by Börje Forssell.
      p. cm.
    Includes bibliographical references and index.
    ISBN 0-13-751058-6 (case)
    1. Radio in navigation. I. Title.
    VK560.F67   1991
    623.89′32–dc20   90-38463
                                                     CIP

British Library Cataloguing in Publication Data

Forssell, Börje, 1939–
    Radionavigation systems.
    1. Boats & ships. Electronic navigation equipment
    I. Title
    623.893

    ISBN 0–13–751058–6

---

1 2 3 4 5   95 94 93 92 91

Contents

Part II: Satellite systems   193

Preface

Radionavigation as a subject at university level is of interest to several categories of people. The reason for this is the growing dependence on accurate navigational and positioning aids, not only at sea and in the air but also on land. Increased activities offshore by oil companies with articulated requirements and great financial capabilities have contributed to this development. Satellite systems with global and time-continuous coverage and unprecedented accuracy will soon be a reality, thus revolutionizing the navigational and positioning state-of-the-art on a broad scale. At the same time the efficiency, accuracy, coverage and reliability of terrestrial radionavigation systems are continuously improved. Due to the development of technology, both hardware and software, user equipment with vastly improved performance is available at affordable cost.

This book is intended to serve as a text for final-year undergraduates and for postgraduates. It is also intended as a source of information for engineers, geodesists, surveyors, navigators, teachers, etc., who have an interest in radionavigation principles and systems.

The background required to use this book is an undergraduate level of mathematics and mathematical statistics and a basic knowledge of electronics components and systems, especially communication systems. Although the book contains a special chapter concerning relevant aspects of electromagnetic wave propagation, fundamental knowledge of waves, antennae and propagation would prove useful.

The book is divided into two basic parts: terrestrial and satellite systems. The first chapters of each part contain fundamentals of navigation and positioning, e.g. characteristics of the earth, navigation principles, calculation of errors in navigation and positioning, and geometric influence on accuracy. The following chapters contain system descriptions.

In order to keep the presentation within reasonable limits, a number of matters have had to be omitted, among them descriptions of a number of short- and medium-range terrestrial systems, mainly used for positioning.

The list of references given at the end of the book is by no means exhaustive. It is merely intended to show the interested reader where more information can be found, particularly the types of publications dealing with navigational matters. Although the great majority of references are in English there are a few exceptions, indicating that much valuable information is published in other languages. The Norwegian references indicate what subjects in this field have been, or are being, treated in Norway and by myself.

The contents of this book reflect the main content of courses in navigation and positioning given at the Department of Electrical Engineering and Computer Science of the Norwegian Institute of Technology (NTH), University of Trondheim. These courses also deal with other terrestrial radionavigation systems, vessel and aircraft traffic control, radar and inertial navigation systems. It is the hope of the author that this book can be used at other universities for similar courses.

I am most grateful to Kari B. Øien at our Division of Telecommunications who typed the manuscript and gave much valuable help during its preparation. Without her assistance this book would not have been possible. My sincere thanks also to my friend George Preiss and his wife who corrected my English.

Börje Forssell
Trondheim, 1990

Part I

Terrestrial systems

1 The fundamentals of terrestrial navigation

1.1 The shape of the earth

In many contexts, even in navigation, it is sufficient to regard the earth as a sphere. However, it has been known for several hundred years that the spherical shape is only an approximation. Many mathematical descriptions have been suggested for the earth as a geometrical body, and these models have been more and more refined by means of better measurement methods and calculations. The spherical shape may be regarded as the first member of a series expansion, and the number of terms necessary depends on the aim of the description. If the second term of the series expansion is included, the earth is regarded as a rotational ellipsoid, i.e. the body which is formed by rotating an ellipse about its short axis (Figure 1.1). The

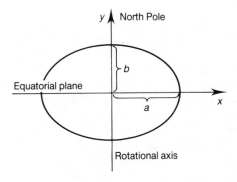

Figure 1.1 A section of the earth as a rotational ellipsoid

usual mathematical description of an ellipse is given in Equation (1.1).

$$\frac{x^2}{a^2} + \frac{y^2}{b^2} = 1 \tag{1.1}$$

The deviation of the ellipse from circular can be characterized by means of the eccentricity, e, or flattening, f, where

$$e = \sqrt{a^2 - b^2}/a \tag{1.2}$$

and

$$f = (a - b)/a \tag{1.3}$$

respectively, thus implying that

$$e = \sqrt{1 - (1 - f)^2} \tag{1.4}$$

The rotational ellipsoid is also an approximation of the real shape of the earth; and it has been and still is difficult to determine this shape with accuracy. Consequently, there exist a number of values of the above parameters. A good approximation is given by $a = 6378.3$ km and $f = 1/298$, i.e. $e = 0.0819$ (see Section 1.2). (When referring to the shape and size of the earth, one means the shape and size of the geoid (see below).)

In fact the cross-section of the earth is shaped somewhat like a pear, and a third-order function, instead of the second-order function given by the ellipse, improves the mathematical description. However, there are still differences between this model and the shape of the earth. One such difference is that the earth is not exactly rotationally symmetric.

There are also additional parameters that are of interest in navigation and positioning, among them the size and direction of the local gravitation. If the direction is used to define a surface which is perpendicular to the direction of gravity and which connects all the points at which the gravity potential is constant and corresponds to an average sea level, a reference equipotential surface called the geoid is obtained (Figure 1.2). This represents a still better approximation of the shape of the earth. At any point on or in the neighbourhood of the surface of the earth, the height above the ellipsoid and/or the geoid can be determined. The difference between the heights of the ellipsoid and the geoid is usually within a few tens of metres.

The mass of the earth is not evenly distributed over its volume;

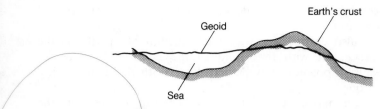

Figure 1.2 The geoid is the averaged sea level

therefore the direction of gravity varies irregularly (Figure 1.3). For surveying and accurate astronomical observations, instruments are used which are levelled so that the vertical axis falls along the direction of gravity. Maps are based on the mathematical representation of the earth, i.e. the ellipsoid.

As astronomical positioning is carried out on the earth's surface, i.e. not usually on the surface of the ellipsoid, and as the vertical is defined as the direction of gravity, which is usually not perpendicular

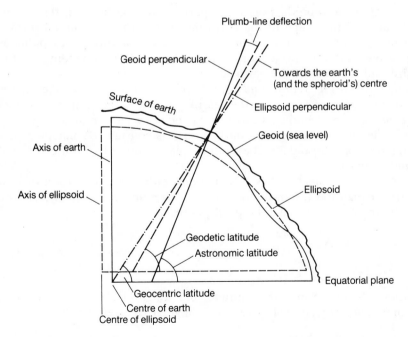

Figure 1.3 Deflection of the plumb-line, geocentric, geodetic and astronomic latitude at a point on the surface of the earth

to the ellipsoid, there may be a deviation between the astronomical and the map positions.

Surveying is also carried out with the direction of gravity as the vertical reference but with the geoid as the horizontal reference. Thus, the resulting position may deviate from both the astronomical and map positions.

Figure 1.3 summarizes the definitions. The angle between the equatorial plane and the direction from the earth's centre to a point on the earth's surface is the *geocentric* latitude of that point. The angle between the equatorial plane and the direction of gravity at the point is the corresponding *astronomic* latitude. Finally, the angle between the equatorial plane and the perpendicular to the ellipsoid is called the *geodetic* latitude. The deflection of the plumb-line is the angle between the direction of gravity and the ellipsoid perpendicular.

1.2 Maps and coordinate systems/Datum

A map is a reproduction of a part of the surface of the earth. Because of the shape of the earth this means that the curved surface is mapped onto a plane. In order to achieve as accurate a mapping as possible, different projection methods are utilized depending on the size and the part of the earth to be mapped. The maps can be either angularly correct (*conformal* projection), i.e. any angle on the map is exactly the same as on the earth, or the area can be correct (*equivalent* projection), i.e. the ratio between any area on the map and the corresponding area on the earth is constant. A mapping of a spherical or ellipsoidal surface onto a plane cannot be both conformal and equivalent.

Any map projection has its equations, either in closed form or as a series expansion (Figure 1.4). With a Cartesian coordinate system the mapping equations have the principal form

$$x = f_1(\Phi, \Lambda)$$
$$y = f_2(\Phi, \Lambda)$$

$$(1.5)$$

where Φ and Λ are the latitude and longitude, respectively, of a point. The projection (Figure 1.4) is angularly correct if

$$\frac{dy}{dx} = \frac{dv}{du}$$

$$(1.6)$$

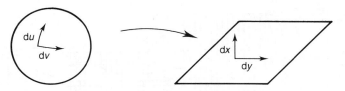

Figure 1.4 Mapping of the earth

and the area is correct if

$$dy \cdot dx = du \cdot dv \qquad (1.7)$$

(u and v are supposed to form a Cartesian system).

The different map projections utilize a projection surface touching the earth at one point or along a curve, or a point of projection, or both. The projections can be classified as *direct* or *indirect*. The first class is characterized by the existence of a geometrically fixed centre of projection (as shown in Figure 1.5). The second class is characterized either by a geometrically defined mapping (but without a fixed centre of projection), or by a mapping after a mathematical transformation. Mercator's projection (see Figure 1.6) is an example of indirect projection.

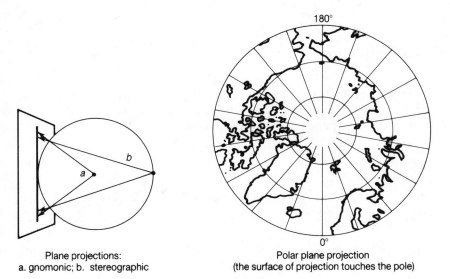

Plane projections:
a. gnomonic; b. stereographic

Polar plane projection
(the surface of projection touches the pole)

Figure 1.5 Plane projection

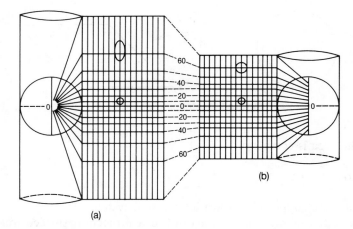

Figure 1.6 Cylindrical projection; **(a)** central cylindrical projection; **(b)** Mercator's conformal projection. The circle at 10° latitude has the same real size as the circle (mapped as an oval) at 60°.

The angular cylindrical projection is very common, especially that of Mercator (Figure 1.6). If the earth is simply regarded as a sphere, Figure 1.4 gives

$$du = R\,d\Phi$$
$$dv = R\cos\Phi\,d\Lambda \tag{1.8}$$

in the longitudinal and latitudinal directions, respectively, where R is the radius of the earth. If all meridians are to be reproduced as straight lines perpendicular to the equator,

$$dy = k\,d\Lambda \tag{1.9}$$

where k is a scale factor. Then according to Equation (1.6) the mapping is angularly correct if

$$dx = \frac{du}{dv}\,dy = \frac{k}{\cos\Phi}\,d\Phi \tag{1.10}$$

Integration gives

$$x = k \ln \left| \tan \left(\frac{\Phi}{2} + \frac{\pi}{4} \right) \right|$$

$$y = k\Lambda$$

(1.11)

Using Equations (1.9) and (1.10) and the definition in Equation (1.7), we see that this projection is only correct with regard to the area along the tangent (where $\Phi = 0$), which explains the large magnification of the polar areas when the equator is the tangent as in this case. For this reason a cylindrical projection where the tangent is a meridian (Figure 1.7) is more suitable for these areas. Such a projection is the Universal Transversal Mercator (UTM), a conformal projection, but reduced in scale by a factor of 0.9996. This means that the cylinder does not touch the earth but intersects, so that there are two close intersection circles which are mapped in true length. The coordinates in the plane are denoted N, E (i.e. north and east). UTM is standardized and divides the east–west direction of the earth into sixty zones of 6°, each with a rising number in the east direction (Figure 1.8). In the north direction the zones are designated by letters. Each zone is divided into smaller 'squares' (Figure 1.8(b)) where the coordinates are given in metres referred to the left and

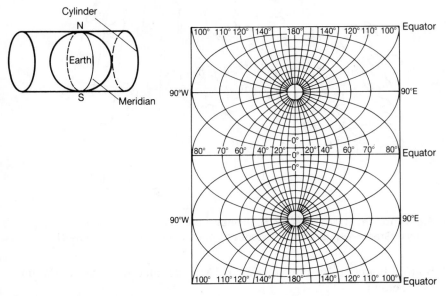

Figure 1.7 A projection cylinder touching the earth along a meridian

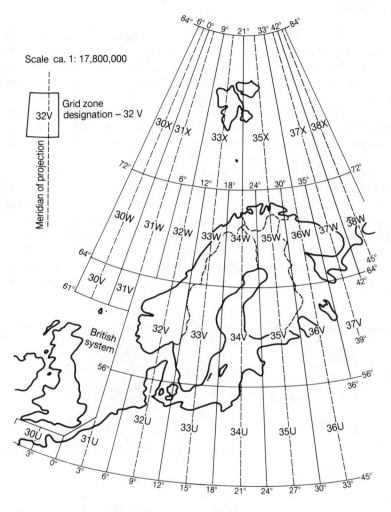

UTM-ZONES SCANDINAVIA – SVALBARD

32V Extended west to 3° E Greenwich

SVALBARD: Extended zones between 72° N and 84° N

Figure 1.8 (**a**) UTM zones for Scandinavia and Svalbard. Extended west to 3° east Greenwich. Svalbard: Extended zones between 72°N and 84°N.

bottom boundary of the 'square'. UTM is coupled to a certain datum (see below).

Formulas (1.11) for the Mercator projection are derived by assum-

100,000 metre squares zone 32V

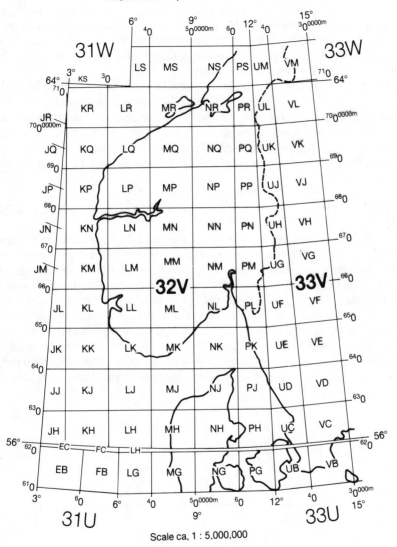

(b) 100 km squares in Zone 32V

Scale ca, 1 : 5,000,000

ing that the earth is a sphere; and they are therefore inaccurate. In real maps, serial expansion methods are used for the parameters to take care of the deviation from the spherical shape. One of the reasons for the popularity of the Mercator maps is the angular correctness such that the course from a point, A, to another point, B, is directly obtainable from the map. As shown in Figure 1.6, even a trip from A to B with a constant course will give a straight course line on the map (a so-called *loxodrome* course). However, this is not the shortest distance between the points.

The coordinate system most frequently used to describe the position of points on or near the earth's surface is based on polar coordinates and a rotational ellipsoid. The meridians are defined with the Greenwich meridian as a reference line, while the equatorial plane is the latitude reference. The relation between geodetic and geocentric latitudes is given by Figure 1.9, Equations (1.1), (1.2) and (A1.3):

$$\tan \Phi_1 = (1 - e^2) \tan \Phi \tag{1.12}$$

Because of the definition, the longitudes and latitudes are dependent on the shape of the ellipse, its size and orientation. These parameters are, therefore, together with a few others, denoted as *datum*. A datum can be global or local. Its main parameters are the x, y and z coordinates of the centre of the ellipsoid, the flattening and the length of the great semi-axis. A certain datum thus characterizes a certain rotational ellipsoid, and the parameters are chosen such that the ellipsoid and the geoid are as close as possible within the area to be considered (the sum of the squares of the deviations is minimized). The most common datum in Western Europe is the European Datum

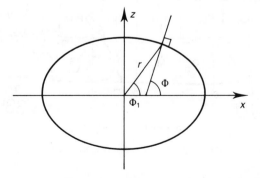

Figure 1.9 Definition of geocentric (Φ_1) and geodetic (Φ) latitudes from the cross-section of the earth as a rotational ellipsoid. As seen, Φ is always greater than Φ_1, but because of the small eccentricity the difference is small (the maximum difference (at about 45° latitude) is about 0.2°)

ED 1950 (the UTM maps in Figure 1.8 are given in ED 1950). As the geoid is not an exact rotational ellipsoid, the matching can be done more accurately for a smaller area. A few examples of datum are given in Table 1.1. Equations of (a simple) transformation from one datum to another are given in Appendix 1. A regional datum is defined by the following eight parameters:

a = great semi-axis of the reference ellipsoid;

f = flattening of the reference ellipsoid;

Φ_0 = latitude of the fundamental point;

Λ_0 = longitude of the fundamental point;

ξ_0 = north–south plumb-line deflection at the fundamental point;

η_0 = east–west plumb-line deflection at the fundamental point;

α_0 = geodetic azimuth at the fundamental point for one side in the triangular network;

N_0 = height of the geoid at the fundamental point (i.e. the distance from the geoid to the reference ellipsoid).

1.3 Distances and directions on the surface of the earth

In many contexts a calculation of the distance and the direction from point A to point B on the earth's surface is needed. The requirements for accuracy of calculation can differ, but an idea of the size of the error is often important.

1.3.1 Great circle course on a spherical earth

In the simplest case the earth is treated as a mathematical sphere. In this case it is well known that the great circle, i.e. the intersection of the spherical surface and the plane containing the two points A and B and the centre of the sphere, gives the shortest path between A and B. (This is based on the fact that any plane intersecting a sphere gives a circular curve of intersection. It is a simple mathematical task to prove that the great circle gives the shortest distance.)

For every calculation of distance and angle on a spherical surface, it is practical to use formulas from spherical trigonometry. Definitions

Table 1.1 Datum examples

Type	Name	Reference ellipsoid	Fundamental point
Regional	European Datum ED 1950	Hayford's ellipsoid or the International Ellipsoid $a = 6378388$ m $f = 1/297$	Helmert's tower, Potsdam (Germany)
Global	World Geodetic System WGS-72	$a = 6378135$ m $f = 1/298.26$	
Global	World Geodetic System WGS-84 (now used by GPS, TRANSIT and LORAN-C)	$a = 6378137$ m $f = 1/298.257223563$ $e = 0.0818191908426$ $e^2 = 0.00669437999013$	

and the most important of these formulas are given in Appendix 2. In order to calculate the great circle course and distance from A to B the formulas (A2.7) and (A2.8) in Section A2.2 can be used. From Figure 1.10 we obtain

$$u = \arccos\left[\sin \Phi_A \sin \Phi_B + \cos \Phi_A \cos \Phi_B \cos (\Lambda_A - \Lambda_B)\right] \qquad (1.13)$$

and

$$\frac{\cos \Phi_B}{\sin \alpha} = \frac{\sin u}{\sin (\Lambda_A - \Lambda_B)} \qquad (1.14)$$

The distances are obtained from Equation (1.13) by multiplying the centre angle, u, by the radius of the earth (= 6378 km); and the course angle, α, related to geographical north is given by Equation (1.14).

1.3.2 Loxodrome course

The loxodrome course is particularly simple to use for Mercator maps when the equator is the tangent (Figure 1.6) because a constant course angle implies a straight line on the map. (Sea maps in Mercator projection are still the most common but are usually not used for latitudes above 70°.)

For loxodrome course calculations it is assumed that the earth is a sphere. A surface element on the northern hemisphere (Figure 1.11) gives

$$\tan \alpha = \frac{a \cos \Phi \, d\Lambda}{a \, d\Phi} \qquad (1.15)$$

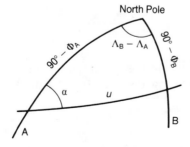

Figure 1.10 Great circle course from A (latitude Φ_A, longitude Λ_A) to B (latitude Φ_B, longitude Λ_B); u is the aspect angle of the great circle path from A to B seen from the earth's centre

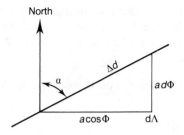

Figure 1.11 Loxodromic course

and by integration

$$\int_A^B d\Lambda = \Lambda_B - \Lambda_A = \tan\alpha \int_A^B \frac{d\Phi}{\cos\Phi} = \tan\alpha \ln\left|\frac{\tan\left(\dfrac{\Phi_B}{2} - \dfrac{\pi}{4}\right)}{\tan\left(\dfrac{\Phi_A}{2} - \dfrac{\pi}{4}\right)}\right|$$

$$(1.16)$$

Thus the course angle is

$$\alpha = \arctan\left[\frac{\Lambda_B - \Lambda_A}{\ln\left|\dfrac{\tan((\Phi_B/2) - (\pi/4))}{\tan((\Phi_A/2) - (\pi/4))}\right|}\right] \qquad (1.17)$$

The distance from A to B along this course line is correspondingly given in Figure 1.11

$$d = \int_A^B \frac{a}{\cos\alpha} d\Phi = \frac{a(\Phi_B - \Phi_A)}{\cos\alpha} \qquad (1.18)$$

The difference between the great circle and loxodromic distances increases with increasing distance from A to B and with increasing latitude.

1.3.3 Distances and directions on an ellipsoidal earth

By definition the length of a curve on a surface is

$$l = \int_A^B R_c(\varphi)\,d\varphi \qquad (1.19)$$

where R_c is the radius of curvature and φ is the aspect angle along the curve. According to Appendix 3 the meridional radius of curvature is

$$M = \frac{a(1 - e^2)}{(1 - e^2 \sin^2 \Phi)^{3/2}} \qquad (1.20)$$

Consequently, the north–south distance between latitudes Φ_A and Φ_B is

$$l_{AB} = \int_{\Phi_A}^{\Phi_B} \frac{a(1 - e^2)}{(1 - e^2 \sin^2 \Phi)^{3/2}}\,d\Phi \qquad (1.21)$$

This elliptical integral cannot be solved explicitly, but utilizing the serial expansion

$$(1 - e^2 \sin^2 \Phi)^{-3/2} = 1 + \tfrac{3}{2}e^2 \sin^2 \Phi + \tfrac{3}{2} \times \tfrac{5}{4}e^4 \sin^4 \phi + \ldots \qquad (1.22)$$

and the relations

$$\sin^2 \Phi = \tfrac{1}{2}(1 - \cos 2\Phi)$$
$$\sin^4 \Phi = \tfrac{3}{8} - \tfrac{1}{2}\cos 2\Phi + \tfrac{1}{8}\cos 4\Phi \qquad (1.23)$$
$$\vdots$$

the integral can be solved.

For arbitrary positions of the points A and B on the rotational ellipsoid the calculations are more complicated. The reason is that the surface perpendiculars at the two points are skew to each other when the points do not have the same latitude or longitude (Figure 1.12). This means that the vertical plane at A containing the point B is not identical to the vertical plane at B containing the point A. Thus the curves from A to B and from B to A usually differ by a very small amount. The distances differ, too, for the same reason. As indicated by Equations (1.21)–(1.23), approximations have to be made in any case, and in general an average of the A and B values can be used. The interested reader is referred to Thomas (1968) for details. For practical cases when the spherical approximation (Equations

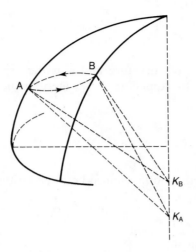

Figure 1.12 The surface perpendiculars of two arbitrary points on a rotational ellipsoid intersect the symmetry axis at different points.

(1.13)–(1.14)) is not accurate enough, computer programs or subroutines using elliptic approximations are available.

2 Error calculations

2.1 Lines of position

Use of navigation or positioning systems in a plane, with reference to known transmitter positions, always gives so-called lines of position (LOP), but surfaces of position in three-dimensional navigation. These lines differ, depending on what measurement principles are used in the specific system. Every navigation system gives information on angles and/or distances, and, in principle, different combinations of information give rise to four different types of intersection of the lines of position.

If the distance between a receiver and a point of reference is measured, the information implies that the receiver is on a circle with the point of reference at the centre and the measured distance as the radius. If the direction from the receiver to the point is measured, the receiver is on a straight line in the measured direction from the point of reference. If the difference in distance to two given points is measured, the receiver is on a hyperbola, the focii of which are the two given points. Two such lines of position then have to intersect in order that a position can be defined (Figure 2.1).

2.1.1 Errors in lines of position

Any measurement contains uncertainty and has statistical errors following a certain (but usually unknown) probability distribution. In addition there are systematic errors, sometimes dominating, which might be eliminated by calibration. An example of systematic errors

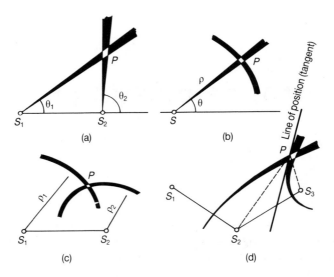

Figure 2.1 Different intersections of lines of positions: (**a**) two angle measurements; (**b**) distance and angle measurement; (**c**) two distance measurements; (**d**) measurement of two distance differences

is an error in the assumed signal propagation velocity (see next chapter). Systematic errors may vary, too, e.g. as a function of time and position. Thus, to some extent it is a matter of definition whether an error type is termed statistical or systematic. Because of such errors a corresponding change appears in the line of position. The displacement of the lines of position in the different cases is indicated in Figure 2.1. An angular error changes the direction of the corresponding line of position at an angular measurement (Figure 2.1(a) and (b)), while a distance error gives a parallel displacement of the corresponding line of position (Figure 2.1(b) and (c)). There is also a parallel displacement in the case of the hyperbola if the transmitters are at a great distance, which they normally are. It can be proved from the definition of the hyperbola that the tangent at any point on the hyperbola divides the angle between the directions to the two focii into two equal parts (Figure 2.1(d)). A measurement error and corresponding errors of position do not change the directions to the transmitters if these are at a greater distance, and do not change the direction of the dividing line of the aspect angle either. So this line is still a tangent of the hyperbola.

The measurement errors have many different causes and any of these causes can have different non-Gaussian probability distributions. Because of the central limit theorem, the total error has a distribution which could be regarded as Gaussian with good approximation. Every random measurement error is therefore regarded as having a Gaussian distribution with a mean of zero and a density function of the type

$$p(\alpha) = \frac{1}{\sqrt{2\pi}\,\sigma} \exp\left(-\frac{\alpha^2}{2\sigma^2}\right) \tag{2.1}$$

(A non-zero mean might be indicative of a systematic error.)

2.2 The geometrical influence on two-dimensional position errors

2.2.1 Errors in measurements of distance and distance differences

2.2.1.1 Distance measurements

In order to measure a position in a plane by means of distance measurements, two points of reference are needed (Figure 2.1(c)), and it is necessary that ambiguity (the circles intersect at two points) does not create any problems. Suppose that the measurement errors in the two directions correspond to distance errors with standard

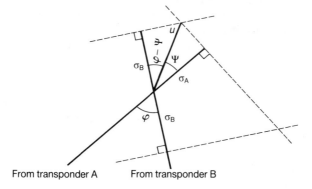

Figure 2.2 Displacement of lines of position at two measurements of distance

deviations σ_A and σ_B (Figure 2.2). From the diagram we obtain

$$u = \frac{\sigma_B}{\cos(\varphi - \Psi)} = \frac{\sigma_A}{\cos \Psi} \qquad (2.2)$$

Equation (2.2) gives the relation between φ and Ψ:

$$\tan \Psi = \left(\frac{\sigma_B}{\sigma_A} - \cos \varphi\right)\Big/\sin \varphi \qquad (2.3)$$

and consequently

$$u = \frac{\sigma_A}{\sin \varphi} \sqrt{1 + \left(\frac{\sigma_B}{\sigma_A}\right)^2 - \frac{2\sigma_B}{\sigma_A} \cos \varphi} \qquad (2.4)$$

(In all the relations illustrated in this and the following sections, it is assumed that the errors of position are small so that the earth within the area can be regarded as a plane surface.)

Figure 2.2 shows that the geometrical problem has two solutions as the displacement of the lines of position can have two directions. If the B line of position is displaced downwards in the diagram, φ has to be replaced by its complementary angle, $\pi - \varphi$, and the result is then

$$u_1 = \frac{\sigma_A}{\sin \varphi} \sqrt{1 + \left(\frac{\sigma_B}{\sigma_A}\right)^2 + \frac{2\sigma_B}{\sigma_A} \cos \varphi} \qquad (2.5)$$

The quantities u and u_1 give the largest position error irrespective of direction if the measurement errors are kept within $\pm \sigma_A$ and $\pm \sigma_B$, respectively, for the two distance mesurements. In many cases $\sigma_A = \sigma_B$ and the result is then

$$u = \frac{\sigma_A}{\cos \dfrac{\varphi}{2}} \qquad (2.6)$$

and

$$u_1 = \frac{\sigma_A}{\sin \dfrac{\varphi}{2}} \qquad (2.7)$$

which can be combined into

$$u = \begin{cases} \dfrac{\sigma_A}{\cos\dfrac{\varphi}{2}}, & 90° < \varphi \le 180° \\[2em] \dfrac{\sigma_B}{\sin\dfrac{\varphi}{2}}, & 0° < \varphi \le 90° \end{cases} \qquad (2.8)$$

If it is assumed for a certain measurement that the errors are d_A and d_B in the A and B directions, respectively, Equations (2.2)–(2.4) can be used to calculate the corresponding value of u. By squaring this value of u and taking the resulting average as $\sigma_A^2 = \overline{d_A^2}$ and $\sigma_B^2 = \overline{d_B^2}$

$$\overline{u^2} = \frac{1}{\sin^2 \varphi} (\sigma_A^2 + \sigma_B^2) \qquad (2.4a)$$

as it is assumed that the errors in the A and B directions are uncorrelated (i.e. $\overline{d_A d_B} = 0$).

Consequently, the root mean square position error is

$$\sqrt{\sigma_A^2 + \sigma_B^2}/\sin \varphi, \text{ or } \sqrt{2}\,\sigma_A/\sin \varphi \text{ if } \sigma_B = \sigma_A$$

(Note that this error is equal to the maximum position error if both the A and B directions give distance errors equal to the standard deviation when the angle between the directions is optimum.)

The locus of a certain position error is given by Equation (2.8) and is a circle through the two points of reference A and B to which the distances are measured (Figure 2.3). The radius of the circle is given by the magnitude of the error. (The locii are circles because an angle based at two points on the periphery, a so-called *periphery angle*, does not change when the point follows the circle.) The minimum error is given by the aspect angle of 90° and is $\sigma_A \sqrt{2}$. Twice that error is marked in Figure 2.3: the results when $\varphi = 41°$ and 139°.

2.2.1.2 Measurements of distance differences

Three points of reference are needed to determine the position by means of distance difference measurements, and the result is then two intersecting hyperbolas. Many navigation systems can be used in circular as well as hyperbolic modes.

The assumed configuration of a hyperbolic system with three

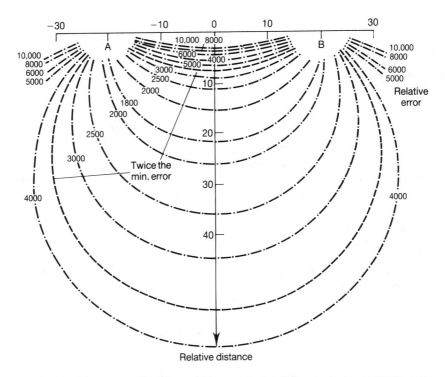

Figure 2.3 Locus of scalar position errors in distance measurements

transmitters is shown in Figure 2.4. In the receiver the distance differences between A and B and between B and C are measured.

As stated earlier, a hyperbola has the property that the tangent at an arbitrary point bisects the angle formed by the directions from this point to the two focii. The angles between a line of position in a hyperbolic navigation system and the directions to the two transmitting stations corresponding to the line of position AB and BC, corresponding to the transmitting combinations A–B and B–C, thus bisect the angles α and β respectively. The angle γ between the lines of position is then $(\alpha + \beta)/2$.

A small error in the measurement of distance differences leads to a displacement of the line of position. The new line of position is parallel to the original one because the directions to the transmitters do not change when the latter are far away (reception of plane wave).

In calculations of the magnitude of the displacement of lines of position at a given error in the distance differences, the configuration is then as shown in Figure 2.5. Here the lines of position are stippled

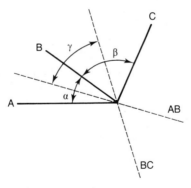

Figure 2.4 Reception geometry in the case of three transmitting stations

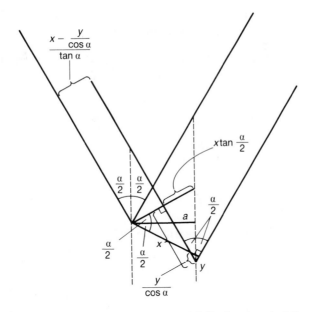

Figure 2.5 Measurement errors imply parallel displacement of the directions to the stations (large distance, plane wave)

(the erroneous position is to the right in the diagram), and the distance error is divided into errors in distances to each of the two transmitters.

From Figure 2.5 we obtain

$$\frac{x - y/\cos \alpha}{\tan \alpha} + x \tan \frac{\alpha}{2} = \frac{a}{\cos \dfrac{\alpha}{2}}$$

(2.9)

Simplified and with the angles as shown in Figure 2.4 we obtain

$$a = \frac{x - y}{2 \sin \dfrac{\alpha}{2}}$$

(2.10)

$$b = \frac{y - z}{2 \sin \dfrac{\beta}{2}}$$

where x is the distance error to A, y the distance error to B and z the error to C. The distance errors of the two measurements are thus $x - y$ and $y - z$ respectively.

In Figure 2.6 (which in fact is the same as Figure 2.5 but with the errors of both lines of position drawn) we obtain

$$\overline{OP} = b/\sin \gamma$$

(2.11)

$$\overline{PQ} = a/\sin \gamma$$

and, according to the cosine theorem,

$$R^2 = \overline{OP}^2 + \overline{PQ}^2 - 2\,\overline{OP}\,\overline{PQ} \cos \gamma$$

(2.12)

Equations (2.10) and (2.11) inserted into (2.12) give

$$R^2 = \frac{1}{4 \sin^2 \gamma} \left[\frac{x^2 - 2xy + y^2}{\sin^2 \dfrac{\alpha}{2}} + \frac{y^2 - 2yz + z^2}{\sin^2 \dfrac{\beta}{2}} \right.$$

$$\left. - 2(xy - y^2 - xy + yz) \frac{\cos \gamma}{\sin \dfrac{\alpha}{2} \sin \dfrac{\beta}{2}} \right]$$

(2.13)

If the errors x, y and z are statistically independent, the statistical average of xy, xz and yz is 0. Then the mean square error in every

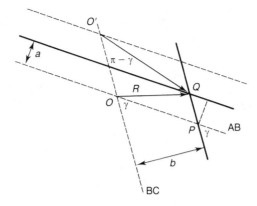

Figure 2.6 The position error is given by a vectorial addition of the two displacements of the lines of position

measurement is

$$\sigma^2 = \overline{(x-y)^2} = \overline{(y-z)^2} \tag{2.14}$$

i.e., if the errors in all directions are equal

$$\frac{\sigma^2}{2} = \overline{x^2} = \overline{y^2} = \overline{z^2} \tag{2.14a}$$

From this we obtain the mean square distance error (position error) R:

$$\overline{R^2} = \frac{\sigma^2}{4\sin^2\gamma}\left(\frac{1}{\sin^2\dfrac{\alpha}{2}} + \frac{1}{\sin^2\dfrac{\beta}{2}} + \frac{\cos\gamma}{\sin\dfrac{\alpha}{2}\sin\dfrac{\beta}{2}}\right) \tag{2.15}$$

If four transmitting stations have been used (and consequently the B station is not in common) the correlation in the third term of Equation (2.13) will disappear (i.e. there will be no quadratic term), and the mean of R^2 will then be

$$\overline{R^2} = \frac{\sigma^2}{4\sin^2\gamma}\left(\frac{1}{\sin^2\dfrac{\alpha}{2}} + \frac{1}{\sin^2\dfrac{\beta}{2}}\right) \tag{2.16}$$

(Note that in such a case, usually, $\gamma \neq (\alpha + \beta)/2$.) The minimum of $\overline{R^2}$ according to Equation (2.16) is $\sigma^2/2$. If this value is used for normalization, we obtain

$$G = \frac{1}{\sqrt{2}\sin\gamma}\left(\frac{1}{\sin^2\dfrac{\alpha}{2}} + \frac{1}{\sin^2\dfrac{\beta}{2}} + \frac{\cos\gamma}{\sin\dfrac{\alpha}{2}\sin\dfrac{\beta}{2}}\right)^{1/2} \qquad (2.17)$$

which is the reduction of accuracy in position determination stemming from the geometrical relations. With four stations the corresponding expression is

$$G\frac{1}{\sqrt{2}\sin\gamma}\left(\frac{1}{\sin 2\dfrac{\alpha}{2}} + \frac{1}{\sin^2\dfrac{\beta}{2}}\right)^{1/2} \qquad (2.18)$$

The minimum is given by $\gamma = \alpha = \beta = 120°$, and in Equation (2.18) the minimum is given by $\alpha = \beta = 180°$ and $\gamma = 90°$. This implies that the optimum location for the three stations is an angularly symmetric location in relation to the receiver, but for four stations the optimum is on the centre line between them, and the centre lines should intersect perpendicularly. G in Equation (2.17) is shown as a function of 2γ and $p = 2\gamma/\alpha$ in Figure 2.7. (Note that the minimum is given by $2\gamma = 240°$ irrespective of the value of p.)

In the same way as in the circular case (Figure 2.2 and Equation (2.8)), γ can be replaced by $\pi - \gamma$ (the point O' in Figure 2.6), and the result is then the change of sign of the third term of Equations (2.12), (2.13), (2.15) and (2.17). In many hyperbolic systems, there is one transmitter (sometimes called the master) to which all distances are referred, and this can have an angular location outside as well as inside the aspect angle of the two other stations. The simplest method of finding the maximum value of R in Equation (2.15) is to select the sign which gives a positive value to the contribution from the third term.

In Equations (2.13)–(2.15) it is assumed that the mean square errors of the two difference measurements are equal. If this is not the case, but the two errors are σ_1 and σ_2 instead, with a cross-correlation

$$k = \overline{(x - y)(y - z)}/\sigma_1\sigma_2 \qquad (2.19)$$

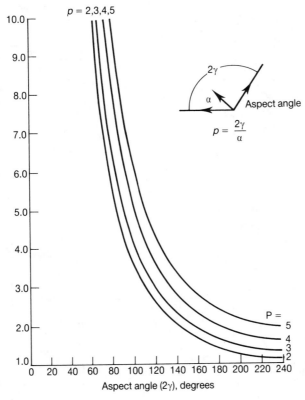

Figure 2.7 Position error as a function of geometry

the result is

$$R = \frac{1}{2 \sin \gamma} \left(\frac{\sigma_1^2}{\sin^2 \frac{\alpha}{2}} + \frac{\sigma_2^2}{\sin^2 \frac{\beta}{2}} - 2k\sigma_1\sigma_2 \frac{\cos \gamma}{\sin \frac{\alpha}{2} \sin \frac{\beta}{2}} \right)^{1/2} \qquad (2.20)$$

When a hyperbolic system consists of a master and more than two other (secondary) stations, the same calculations as given above can be performed for the whole coverage area, using the secondaries which give the best result in the desired direction (i.e. the greatest aspect angle and the least measurement error). For a given accuracy, contours can be drawn which are analogous to the circles in Figure 2.3, but the curves are not usually circular (Equation (2.20)). Some examples of such contours of accuracy are given in Figure 2.8

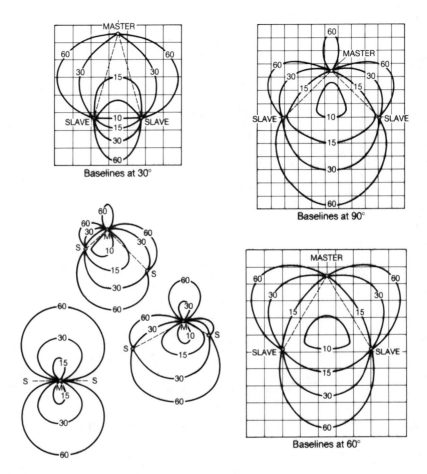

Figure 2.8 Examples of accuracy contours in a hyperbolic system with symmetrical (in a square net) and non-symmetrical station location (courtesy Racal-Decca Marine Navigation, Ltd)

(DECCA, 1979). (The straight lines between the master and the slaves (Decca terminology) are called *baselines*.)

2.2.2 Error ellipses

The derivations of the preceding section apply to scalar position errors, so possible differences in error probability in different directions from the correct point have not been taken into consideration.

In order to consider these, it is necessary to know the probability distribution of the different errors in the lines of position, as well as the geometrical relations.

As mentioned earlier, it is assumed that all measurement results contain errors with an average of zero and a Gaussian distribution. Consequently, this also applies to the displacements of the lines of position caused by the measurement errors. In order to calculate the distribution of the position errors caused by the displacement of two lines of position, it is necessary to find the density function of a quantity depending on two variables, i.e. a two-dimensional Gaussian probability density function. This has the form (Cramér, 1946)

$$p(x, y) = \frac{1}{2\pi\sigma_x\sigma_y\sqrt{1 - \rho_{xy}^2}}$$

$$\times \exp\left[-\frac{1}{2(1 - \rho_{xy}^2)}\left(\frac{x^2}{\sigma_x^2} + \frac{y^2}{\sigma_y^2} - 2\rho_{xy}\frac{xy}{\sigma_x\sigma_y}\right)\right]$$

$$(2.21)$$

where σ_x and σ_y are the standard deviations of the variables x and y, and ρ_{xy} is the correlation between the variables, i.e.

$$\rho_{xy} = \frac{\overline{xy}}{\sigma_x\sigma_y} \qquad (2.22)$$

Equation (2.21) shows that a constant error probability is given by all curves in the xy-plane where

$$\frac{1}{2(1 - \rho_{xy}^2)}\left(\frac{x^2}{\sigma_x^2} + \frac{y^2}{\sigma_y^2} - 2\rho_{xy}\frac{xy}{\sigma_x\sigma_y}\right) = c^2 = \text{const.} \qquad (2.23)$$

Equation (2.23) describes an ellipse, the so-called *error ellipse* which has its centre at the correct position, and whose size and orientation depend on the included coefficients. In order to find the shape of the ellipse a coordinate transformation has to be performed which only implies rotation of the ellipse so that its axes are parallel to the coordinate axes. It then assumes the well-known quadratic shape of Equation (1.1), and the lengths of the axes can be read directly. The probability of being within the ellipse of Equation (2.23) is then $1 - \exp(-c^2)$, which can simply be shown in the following way.

If the rotated ellipse has the shape $x_1^2/\sigma_1^2 + y_1^2/\sigma_2^2 = 1$, the error probability $p_1(x_1, y_1)\,dx_1\,dy_1$ can be integrated over the surface of the ellipse after the substitution of variables $x_1/\sigma_1 \to x_2$, $y_1/\sigma_2 \to y_2$, followed by a transition to polar coordinates r_2, φ_2:

$$\int_{x_1} \int_{y_1} p(x_1 \cdot y_1) \, dx_1 \, dy_1$$

$$= \int_{x_1} \int_{y_1} \frac{1}{2\pi\sigma_1\sigma_2} \exp\left[-\frac{1}{2}\left(\frac{x_1^2}{\sigma_1^2} + \frac{y_1^2}{\sigma_2^2}\right)\right] dx_1 dy_1$$

$$= \int_0^{c\sqrt{2}} r_2 \exp\left(-r_2^2/2\right) dr_2 \int_0^{2\pi} \frac{d\varphi_2}{2\pi} = 1 - \exp\left(-c^2\right)$$

$$(2.24)$$

The calculation of the size of the ellipse for a given error probability and the direction of the axes can be performed in the following way. The ellipse has the general shape

$$x_{11}x^2 + x_{22}y^2 + x_{12}xy = k \tag{2.25}$$

where the coefficients x_{11}, x_{22} and x_{12} and the constant k can be found in Equation (2.23). When $x_{12} \neq 0$, the axes of the ellipse are not parallel to the coordinate axes. In order to find the direction and length of the axes of the ellipse, the ellipse is rotated through an angle ε by means of a coordinate transformation, so that its equation after the transformation contains quadratic terms only (Figure 2.9).

The transformation equations are

$$x = u \cos \varepsilon - v \sin \varepsilon \tag{2.26}$$

$$y = u \sin \varepsilon + v \cos \varepsilon \tag{2.27}$$

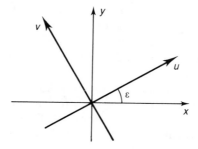

Figure 2.9 Rotation of the coordinate system

Equations (2.26) and (2.27) inserted into Equation (2.25) give

$$x_{11}(u \cos \varepsilon - v \sin \varepsilon)^2 + x_{22}(u \sin \varepsilon + v \cos \varepsilon)^2$$
$$+ x_{12}(u \cos \varepsilon - v \sin \varepsilon)(u \sin \varepsilon + v \cos \varepsilon) = k$$

$$(2.28)$$

and the condition that the coefficient of uv shall be zero gives

$$-2x_{11} \sin \varepsilon \cos \varepsilon + 2x_{22} \sin \varepsilon \cos \varepsilon + x_{12}(\cos^2 \varepsilon - \sin^2 \varepsilon) = 0 \quad (2.29)$$

and consequently

$$\tan 2\varepsilon = \frac{x_{12}}{x_{11} - x_{22}} \tag{2.30}$$

The coefficient of u^2 is (Equation (2.28))

$$u_{11} = x_{11} \cos^2 \varepsilon + x_{22} \sin^2 \varepsilon + x_{12} \cos \varepsilon \sin \varepsilon$$
$$= x_{11} \frac{1 + \cos 2\varepsilon}{2} + x_{12} \frac{\sin 2\varepsilon}{2}$$
$$= \frac{x_{11} + x_{22}}{2} + \frac{1}{2} \cos 2\varepsilon (x_{11} - x_{22} + x_{12} \tan 2\varepsilon) \tag{2.31}$$

Equation (2.30) inserted into Equation (2.31) gives

$$u_{11} = \tfrac{1}{2}(x_{11} + x_{22}) + \tfrac{1}{2}(x_{11} - x_{22}) \cos 2\varepsilon (1 + \tan^2 2\varepsilon)$$
$$= \tfrac{1}{2}(x_{11} + x_{22}) + \tfrac{1}{2} \sqrt{(x_{11} - x_{22})^2 + x_{12}^2} \tag{2.32}$$

Correspondingly, the coefficient of v^2 becomes

$$u_{22} = \tfrac{1}{2}(x_{11} + x_{22}) - \tfrac{1}{2} \sqrt{(x_{11} - x_{22})^2 + x_{12}^2} \tag{2.33}$$

Consequently, the semi-axes of the ellipse have the following lengths:

$$a_1 = \sqrt{k/u_{11}} \tag{2.34}$$

and

$$a_2 = \sqrt{k/u_{22}} \tag{2.35}$$

The above equations contain an ambiguity as the major axis lies either along the u or the v axis. The reason for this is that the tangent function (Equation (2.30)) has a period of π so that the solution of Equation (2.23) has the form

$$\varepsilon = \frac{1}{2} \arctan\left(\frac{x_{12}}{x_{11} - x_{22}}\right) + i \cdot \frac{\pi}{2} \tag{2.36}$$

where i can be either 0 or 1. The ambiguity can, however, be solved in the following way.

Differentiation of Equation (2.25) with regard to x gives

$$x_{11}x + x_{22}y \frac{dy}{dx} + x_{12}\left(y + x \frac{dy}{dx}\right)\Big/2 = 0 \tag{2.37}$$

$$\frac{dy}{dx} = -\frac{x_{11}x + x_{12}y/2}{x_{22}y + x_{12}x/2} \tag{2.37a}$$

On the y-axis (Figure 2.10):

$$\left(\frac{dy}{dx}\right)_{x=0} = -\frac{x_{12}}{2x_{22}} \tag{2.38}$$

As x_{22} is always positive, the major axis of the ellipse lies in the second and fourth quadrants, when $x_{12} > 0$, and in the first and third quadrants when $x_{12} < 0$.

Mathematically the above can be written

$$\varepsilon = \begin{cases} \varepsilon' - 90°, \text{ if } x_{12} \text{ and } \varepsilon' \text{ have the same sign} \\ \varepsilon' \text{ otherwise;} \end{cases} \tag{2.39}$$

where

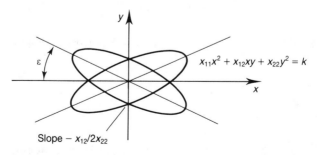

Figure 2.10 Determination of the directions of the major axis of the ellipse

$$\varepsilon' = \frac{1}{2} \arctan \left(\frac{x_{12}}{x_{11} - x_{22}} \right) \tag{2.40}$$

Equations (2.25)–(2.40) describe in detail how the error ellipse is found in the general case. In many cases the computation work can be reduced by a suitable choice of coordinate axes. The physical implications of the ellipses could be summarized in a few points: (1) The larger the area of the ellipse, other parameters held constant, the larger the percentage of measurements giving a position within it. (2) The more one LOP error dominates, the more elongated is the ellipse. (3) The more one LOP error dominates, the closer is the direction of the major axis to the direction of displacement of the LOP.

2.2.2.1 Distance measurement error ellipses

Position determination by means of distance measurement (so-called ρ–ρ navigation), described in Section 2.2.1.1, gives error ellipses which can be calculated directly by using Equations (2.21)–(2.40). In Figure 2.11, l_1 and l_2 are the measured distances and Δx and Δy are the x- and y-coordinates resulting from the measurement error. We then have

$$l_A = \Delta l_1 = \frac{x_P \Delta x + y_P \Delta y}{\sqrt{x_P^2 + y_P^2}} = \frac{x_P \Delta x + y_P \Delta y}{l_1} \tag{2.41}$$

$$l_B = \Delta l_2 = \frac{(x_P - x_B)\Delta x + y_P \Delta y}{\sqrt{(x_P - x_B)^2 + y_P^2}} = \frac{(x_P - x_B)\Delta x + y_P \Delta y}{l_2} \tag{2.42}$$

The measurements of the distances to A and B are mutually independent, and Equation (2.23) can thus be used with $\rho_{xy} = 0$.

$$\frac{l_A^2}{\sigma_A^2} + \frac{l_B^2}{\sigma_B^2} = 2c^2 \tag{2.43}$$

Insertion of Equations (2.41) and (2.42) into Equation (2.43) gives

$$\frac{(x_P \Delta x + y_P \Delta y)^2}{l_1^2 \sigma_A^2} + \frac{[(x_P - x_B)\Delta x + y_P \Delta y]^2}{l_2^2 \sigma_B^2} = 2c^2 \tag{2.44}$$

or

$$x_{11}(\Delta x)^2 + x_{12}\Delta x \Delta y + x_{22}(\Delta y)^2 = 2c^2 \qquad (2.45)$$

where

$$x_{11} = \frac{x_P^2}{l_1^2 \sigma_A^2} + \frac{(x_P - x_B)^2}{l_2^2 \sigma_B^2} \qquad (2.46)$$

$$x_{12} = 2y_P\left(\frac{x_P}{l_1^2 \sigma_A^2} + \frac{x_P - x_B}{l_2^2 \sigma_B^2}\right) \qquad (2.47)$$

$$x_{22} = y_P^2\left(\frac{1}{l_1^2 \sigma_A^2} + \frac{1}{l_2^2 \sigma_B^2}\right) \qquad (2.48)$$

Figure 2.11 also shows that

$$(x_P/l_1)^2 = 1 - (y_P/l_1)^2 \qquad (2.49)$$

$$(x_P - x_B)^2/l_2^2 = 1 - (y_P/l_2)^2 \qquad (2.50)$$

If $\sigma_A = \sigma_B = \sigma$, insertion of Equations (2.49) and (2.50) into Equations (2.46)–(2.48) gives

$$x_{11} = \frac{1}{\sigma^2}\,[2 - y_P^2(l_1^{-2} + l_2^{-2})] \qquad (2.51)$$

$$x_{12} = \frac{2y_P}{\sigma^2}\,[\pm\, l_1^{-1}\sqrt{1 - y_P^2/l_1^2} \pm l_2^{-1}\sqrt{1 - y_P^2/l_2^2}] \qquad (2.52)$$

where the first minus sign is used if $x_P < 0$ and the second one if $x_P < d$ (Figure 2.11)

$$x_{22} = \frac{y_P^2}{\sigma^2}\,[l_1^{-2} + l_2^{-2}] \qquad (2.53)$$

The size and orientation of the ellipse can now be determined by using the values of x_{11}, x_{12} and x_{22} in Equations (2.32)–(2.35) and Equations (2.39)–(2.40) above with $k = 2c^2$. (If, for example, 68.27 per cent of all measurements (1σ value) are to be found within the ellipse, $1 - \exp(-c^2) = 0.6827$, which gives $k = 2.296$. The 2σ value, 95.29 per cent, gives $k = 6.112$.)

Figure 2.12 shows how the orientations and relative sizes of the error ellipses behave as functions of the position related to the two points A and B from which distances are measured.

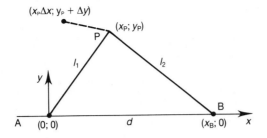

Figure 2.11 Calculation of the error distribution

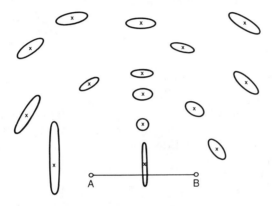

Figure 2.12 Orientation and relative size of the error ellipse

2.2.2.2 Error ellipses for a hyperbolic navigation system

The error ellipses for a hyperbolic system can be calculated in a similar way as for a ρ–ρ system. From Figure 2.13 and Equation (2.10) we have

$$l_1 = \frac{d_1}{2\sin\dfrac{\alpha}{2}} \tag{2.54}$$

$$l_2 = \frac{d_2}{2\sin\dfrac{\beta}{2}} \tag{2.55}$$

where α and β are the angles between the transmitters in pairs.

Figure 2.13 gives

$$x = \frac{l_1 \cos \gamma + l_2}{\sin \gamma} \tag{2.56}$$

$$y = l_1 \tag{2.57}$$

from which

$$l_1 = y \tag{2.58}$$

$$l_2 = x \sin \gamma - y \cos \gamma \tag{2.59}$$

and, after insertion of Equations (2.54) and (2.55),

$$d_1 = 2y \sin \frac{\alpha}{2} \tag{2.60}$$

$$d_2 = 2(x \sin \gamma - y \cos \gamma) \sin \frac{\beta}{2} \tag{2.61}$$

The statistical quantities d_1 and d_2 are usually assumed to have Gaussian distribution, and they then have the joint density function of Equation (2.21). Inserting Equations (2.60) and (2.61) into Equation (2.21) we have

$$p(d_1, d_2) = (2\pi\sigma_1\sigma_2 \sqrt{1 - \rho^2})^{-1}$$

$$\times \exp\left\{\frac{-2}{1 - \rho^2}\left[\frac{y^2 \sin^2 \frac{\alpha}{2}}{\sigma_1^2} + \frac{(x \sin \gamma - y \cos \gamma)^2 \sin^2 \frac{\beta}{2}}{\sigma_2^2}\right.\right.$$

$$\left.\left. - 2\rho\frac{y \sin \frac{\alpha}{2}(x \sin \gamma - y \cos \gamma) \sin \frac{\beta}{2}}{\sigma_1\sigma_2}\right]\right\} \tag{2.62}$$

Now the coefficients of x^2, y^2 and xy are calculated from Equation (2.62) which gives (notation according to Equation (2.25))

$$x_{11} = \frac{\sin \gamma \sin^2 \frac{\beta}{2}}{\sigma_2^2} \tag{2.63}$$

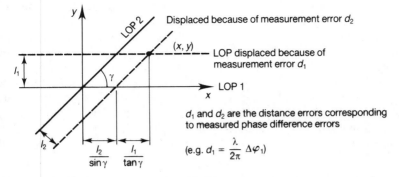

Figure 2.13 Configuration for calculations of error ellipses in a hyperbolic system

$$x_{22} = \frac{\sin^2\frac{\alpha}{2}}{\sigma_1^2} + \frac{\cos^2\gamma\sin^2\frac{\beta}{2}}{\sigma_2^2} + 2\rho\frac{\cos\gamma\sin\frac{\alpha}{2}\sin\frac{\beta}{2}}{\sigma_1\sigma_2} \tag{2.64}$$

$$x_{12} = -\frac{\sin 2\gamma\sin^2\frac{\beta}{2}}{\sigma_2^2} - \frac{2\rho\sin\gamma\sin\frac{\alpha}{2}\sin\frac{\beta}{2}}{\sigma_1\sigma_2} \tag{2.65}$$

$$k = (1 - \rho^2)\frac{c^2}{2} = -\frac{1}{2}(1 - \rho^2)\ln(1 - p) \tag{2.66}$$

where p is the chosen probability to be within the ellipse. The correlation coefficient, ρ, is calculated according to the definition in Equation (2.22):

$$\rho = \frac{\overline{d_1 d_2}}{\sigma_1\sigma_2} \tag{2.67}$$

In order to calculate ρ, the same method as used in Section 2.2.1.2 is used, and the result will therefore be dependent on whether a master is used or not and, if so, how large a part of the error can be ascribed to the master signal. The contribution from the master signal has, in any case, to be smaller than the smallest of the quantities σ_1 and σ_2 so that

$$0 \le |\rho| \le \frac{\min[\sigma_1^2, \sigma_2^2]}{\sigma_1\sigma_2} \tag{2.68}$$

Using Equations (2.63)–(2.67) the error ellipse can be determined by means of Equations (2.32)–(2.35) and (2.39)–(2.40).

Concluding the section on error ellipses, it should be mentioned that the same methods can be used when a position is determined from more than two lines of position. Depending on the different lines of position we obtain different curves for constant error probability, but they are all ellipses. However, calculations become more complicated when the joint density function is to be found; among other things, more correlation functions (of the type ρ_{xy}, ρ_{xz}, ρ_{yz}) have to be calculated.

2.3 Accuracies

Many people who deal in some way with positioning or navigation systems need to know the accuracy of the system. In order that accuracy information is to have some value, additional information is needed on the kind of accuracy. Such additional information is not always given, and, if given, the information may be incomplete, misleading or erroneous. The information may also be given in a way that the user does not understand. This chapter is an attempt to make clear what lies behind the different notions of accuracy and to characterize the relations between them. Only random errors are regarded, and as before they are assumed to be Gauss distributed with zero mean. The lines of position are assumed to be straight over the error area which is two-dimensional.

The simplest notion is probably that of percentages which are related to a certain accuracy contour, i.e. a circle or an ellipse. For example, the figure 75 per cent tells us that three-quarters of all measurements will give results within the drawn curve. However, the figure does not say anything about the distribution of the errors inside and outside the curve, and accurate information on this can only be obtained if many curves with different percentages are drawn beside each other.

Instead of percentages it is common to use the standard deviations of Gaussian distributions which correspond to a certain percentage. (The variance is the square of the standard deviation.) If a random variable has a sigma value or a standard deviation or an rms value (rms = root mean square), this means that 68.3% of all values of the variables are within ± the mentioned figure, 95.4 per cent are within ± twice the figure, 99.6 per cent are within ± three times the figure, etc. A numerical integration of the density function of the Gaussian distribution gives the percentages, i.e.

$$p(x_0) = \int_{-x_0}^{x_0} \frac{1}{\sqrt{2\pi}\,\sigma} \exp\left(-\frac{x^2}{2\sigma^2}\right) dx \tag{2.69}$$

In tables the so-called *error function* is often used:

$$\mathrm{erf}(y_0) = \frac{2}{\sqrt{\pi}} \int_0^{y_0} \exp(-y^2) dy \tag{2.70}$$

and the value of $p(x_0)$ is obtained by using $y_0 = x_0/\sqrt{2}\sigma$ as the argument of erf.

In the two-dimensional error functions used here, the same notations are occasionally used as in one-dimensional functions, especially when the error is directionally independent, i.e. when the error ellipse is a circle. Equation (2.4a) gives the standard deviation in such a situation when the angle between the lines of position is $90°(\varphi = 90°$ in Equation (2.4a)). Note that this standard deviation is then $\sqrt{2}$ times the standard deviation of the single line of position. This is proved in Appendix 4.

Usually the error ellipse has an eccentricity > 0, and then it is absolutely necessary to note the precise definition of the terms. When, for example, an error is denoted only as 'x m (2σ)', it is not possible to know if it is referred to a one-dimensional standard deviation or to a two-dimensional circular standard deviation. That is, if there is 95.4 per cent or 86.4 per cent probability of the error being less than x m. In order that such an uncertainty should not exist, the accuracy must be given instead as 'x m (95 per cent)' or 'x m (86 per cent)', but even then an uncertainty exists, i.e. the directional dependence. In spite of the latter uncertainty it is usual to denote the error as a radius of a circle within which a certain percentage of all measurements should be found. The most common percentage is 50, and the mentioned radius is then called the *circular error probable* (CEP). (A similar definition exists in three dimensions, the *spherical error probable* (SEP).) This can be calculated by integration of density functions of the type given in Equation (2.21) over a circular surface. But as the contours of constant error probability are ellipses, CEP depends on the eccentricity. Different approximations of CEP as functions of the shape of the ellipse exist, some of which are given in Figure 2.14 as functions of the ratio between the semi-minor and semi-major axes of the ellipse (which can be calculated from Equations (2.34) and (2.35)). The CEP value in the diagrams is normalized to the σ value along the major axis, σ_u, which is of course

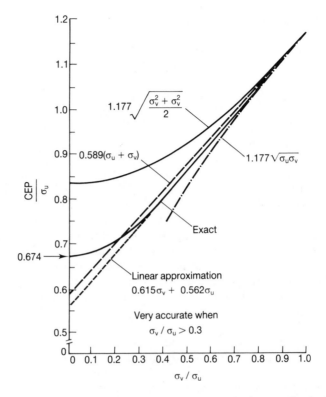

Figure 2.14 CEP and some approximations for elliptical error distributions. The value for $\sigma_v = \sigma_u$ is given by Equation (A4.6) and $c^2 = \ln(1 - p)$, giving $r = \sigma_u \sqrt{-2\ln(1 - p)} = 1.177\sigma_u$. The value for $\sigma_V = 0$ is $\sqrt{2}\,y_{0.5}$, where $\operatorname{erf}(y_{0.5}) = 50\%$

identical to σ_x. (If the reader wishes to check, the integration in v should be performed.)

The area of the surface of the ellipse containing 50 per cent of all errors is always smaller than the corresponding area of the CEP circle. The difference between the areas grows with the eccentricity of the ellipse, which means that the use of CEP is limited to elliptical error functions with a smaller eccentricity (i.e. less than $e = 0.5$ which gives an area ratio of about 0.99). In such cases the similarity of the surface areas can be used to calculate CEP approximately from the size of the 50 per cent ellipse (the area of the ellipse is the product of the semi-axes times π).

Even if CEP is the most used measure of circular accuracy, it may sometimes be desirable to find the radius of percentages other than

50. This can be calculated from Equation (A4.6), and the probability of being within a certain circle is then

$$p_s(r) = 1 - \exp\left(\frac{r^2 \ln 2}{r_{CEP}^2}\right) \tag{2.71}$$

i.e.

$$r(p_s) = r_{CEP} \sqrt{\frac{-\ln(1 - p_s)}{\ln 2}} \tag{2.72}$$

Figure 2.15 indicates the relations between the different radii.

Another common way of denoting accuracy is d_{rms} which is defined (Figure 2.16) as

$$d_{rms} = \sqrt{\sigma_n^2 + \sigma_v^2} \tag{2.73}$$

σ_u and σ_v are defined as the standard deviations of the one-dimensional density functions in the u- and v-directions. Like CEP, d_{rms} is also dependent on the eccentricity of the ellipse, and a circle with a radius $1d_{rms}$ thus contains different percentages of all measurements according to the σ_v/σ_u ratio.

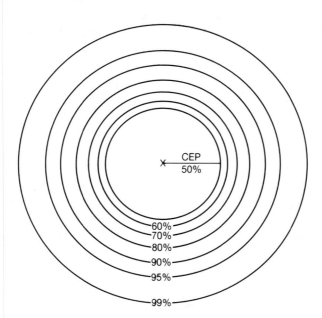

Figure 2.15 Relations between error circles

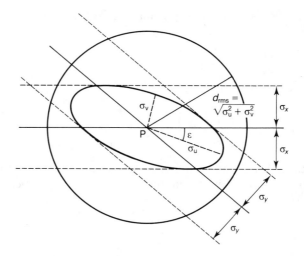

Figure 2.16 Definition of d_{rms}

When σ_v/σ_u varies between 0 and 1, d_{rms} varies between σ_u and $\sigma_u\sqrt{2}$ and at the same time the probability of being within a circle with radius $1d_{rms}$ varies between 63.2 per cent and 68.3 per cent. The corresponding limits of probability for $2d_{rms}$ are 98.2 per cent and 95.4 per cent. The relations between d_{rms} and CEP are shown in Figure 2.17. Remember that d_{rms} is defined as a radius which can give

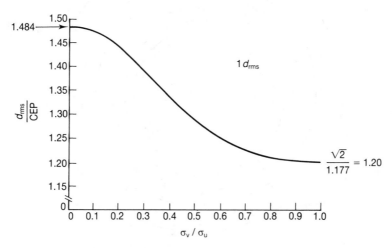

Figure 2.17 Ratio between d_{rms} and CEP

different percentages, while CEP is defined as a percentage which can give different radii.

The percentages are calculated in the following way. When $\sigma_v = 0$, the distribution is one-dimensional, and $1\,d_{rms}$ corresponds to 1σ (i.e. 68.3 per cent), $2\,d_{rms}$ to 2σ (i.e. 95.4 per cent), etc. When $\sigma_v = \sigma_u$, we have a radial Rayleigh distribution (Equation (2.71)), and the probabilities are $1 - \exp[-\frac{1}{2}(\sqrt{2})^2] = 63.2$ per cent and $1 - \exp[-\frac{1}{2}(2\sqrt{2})^2] = 98.2$ per cent respectively. Note the different behaviour of the probability curves for 1 and $2\,d_{rms}$ as functions of σ_v/σ_u because of different distributions.

2.4 Two-dimensional position determination by means of more than two LOPs

An important case in position determination is the situation when more than the minimum number of LOPs is available. The question now is how should the available information be used in order to achieve the best result. Depending on the circumstances, several methods are possible, but probably the most common is the method of least squares. It can be used irrespective of the navigation system. Use of this method implies that the receiver is assumed to be at a certain position. Then the measurement results are compared with the calculated results corresponding to the assumed position. From the difference found, the assumed position is changed such that the result of the calculation coincides with the measured position. As only the first derivative of the difference is taken into consideration at the change, the method is often called linearization. The existence of more than the minimum number of LOPs implies that more measurements are made each time than the number of coordinates to be determined. This gives a system of equations, the solution of which can be optimized. Thus the method could be regarded as just a mathematical means for optimum processing of measurement results, which in fact it is. However, as many navigational receivers use it, the method is worth discussion in this context.

As the mathematical description of the method is more demanding than the mathematics used in the preceding parts of this chapter, the details are given in Appendix 5, where the interested reader can find the applicable formulas.

③ Wave propagation

For all navigation systems based on radio waves a sound knowledge of wave parameters is crucial to enable us to make an accurate position determination. The systems described in the following chapters operate at frequencies between 10 kHz and 10 GHz, and the presentation is therefore limited to this frequency range. As will be seen, there are wide propagation differences.

3.1 Free space propagation

If a transmitter with a power of P_t (we use subscripts t for transmission, r for reception) is connected to a lossless and totally non-directional antenna, i.e. an antenna radiating equal amounts of power in all directions, the power density at distance R is equal to the radiated power divided by the surface area of the sphere with radius R,

$$r_h = \frac{P_t}{4\pi R^2} \tag{3.1}$$

In reality there are no totally non-directional antennae, but every antenna concentrates its radiation in certain directions. This 'concentration ability' is called the antenna *directivity* or *gain* and is often denoted by D_t which thus is a function of the elevation and azimuth directions, i.e. $D_t(\theta, \Phi)$. (This is the gain of the lossless antenna.) The power density at distance R is then

$$r_D = \frac{D_t P_t}{4\pi R^2} \tag{3.2}$$

If losses in the antenna are also considered, e.g. resistive losses in the antenna structure itself, radiation from feeders which does not hit the reflector, reflections in the feeder lines, etc., the effective gain, usually denoted G, is obtained and

$$G_t = kD_t \tag{3.3}$$

where k is usually between 0.3 and 0.8. Similarly to D_t, G_t is also a function of direction, i.e. $G_t(\theta, \Phi)$. The resulting power density is then

$$r_G = \frac{G_t P_t}{4\pi R^2} \tag{3.4}$$

A receiver antenna at this position is assumed to have an effective area A_r, and consequently a gain (Kai Fong Lee, 1984):

$$G_r = \frac{4\pi A_r}{\lambda^2} \tag{3.5}$$

where λ is the wavelength. The received power is then

$$P_m = \frac{G_t P_t A_r}{4\pi R^2} = \frac{G_t G_r^2 \lambda}{(4\pi R)^2} P_t \tag{3.6}$$

The received power can thus be written as the product of the gains of the transmitting and receiving antennae and the transmitter power divided by an attenuation factor:

$$a = \left(\frac{4\pi R}{\lambda}\right)^2 \tag{3.7}$$

This factor is called *space attenuation*. When a communication system is to be designed with regard to, for example, transmitter power based on given distances and antenna parameters, this factor has to be taken into account. Usually, decibel is used for simple and easy calculations, and space attenuation can thus be written

$$a_1 = 20\log\left(\frac{4\pi R}{\lambda}\right) \text{ (dB)} \tag{3.8}$$

Space attenuation increases quadratically with frequency and distance. Given the antenna area the gain also increases quadratically with frequency (Equation (3.5)), within certain limits, and this can be utilized to compensate for the increasing space attenuation. If the two constant antenna areas are A_t and A_m, respectively, Equation (3.6) can be written

$$P_m = \frac{A_t A_m P_t}{(\lambda R)^2} \qquad (3.9)$$

and then the received power actually increases with frequency. Most often, however, in a real antenna losses increase as a function of frequency, so that the effective area is not constant. Equation (3.9), with constant values of A_t and A_m, thus expresses an ideal relation only.

3.2 Reflection from surroundings

There is the constant problem of reflection from terrain or sea, not to mention buildings and other man-made objects.

Calculation of reflection coefficients can usually be performed using Maxwell's equations as in any case of calculation of electric and magnetic fields and their behaviour.

Problems, however, rapidly become unmanageable due to variations in the electric and geometric parameters of the reflective surfaces, and a great number of irregularities which are not always small in comparison to the wavelengths. Therefore a statistical approach is often used in the calculations, to a smaller or greater extent, and a certain terrain area is divided into several smaller areas where such properties as slope and roughness determine the divisions.

A common method is to distinguish between two main types of reflections: specular and diffuse. A *specular reflection* is caused by a smooth (compared to the signal wavelength) surface, comes from a defined direction within the first Fresnel zone and has a defined phase. (A Fresnel zone is the area within which the path length difference between the direct and the reflected wave varies by less than half a wavelength.) *Diffuse reflections* (sometimes called 'scattering') are less directive and arise in a larger area than the specular ones. A diffuse reflection is composed of a very large number of small reflections with comparable magnitudes, having a random phase uniformly distributed between 0 and 2π. Therefore, the resulting magnitude is Rayleigh distributed, i.e. its magnitude density function is

$$P(x) = \frac{2x}{P_0} \exp(-x^2/P_0), \ x > 0 \tag{3.10}$$

where P_0 is the average power.

The relations, in the case of specular reflections are found by means of geometrical optics (angle of incidence = angle of reflection). The reflection coefficient itself is given by Fresnel equations, and for vertical polarization we have (David and Voge, 1969)

$$R_v(\theta) = \frac{\varepsilon \cos \theta - \sqrt{\varepsilon - \sin^2 \theta}}{\varepsilon \cos \theta + \sqrt{\varepsilon - \sin^2 \theta}} \tag{3.11}$$

where θ (Figure 3.3) is the angle of incidence (referred to the surface perpendicular), and $\varepsilon = \varepsilon' - j\varepsilon''$ is the complex relative dielectric constant of the reflective surface. The corresponding formula for horizontal polarization is

$$R_h(\theta) = \frac{\cos \theta - \sqrt{\varepsilon - \sin^2 \theta}}{\cos \theta + \sqrt{\varepsilon - \sin^2 \theta}} \tag{3.12}$$

As examples, Figures 3.1–3.2 show values of R_v and R_h, respectively, for real values of ε (i.e. $\varepsilon'' = 0$, which means no reflection losses). The curves also show similar behaviour if $\varepsilon'' \neq 0$, but then R_v does not reach zero for nearly grazing incidence, the so-called *Brewster angle*. This angle is found from Equation (3.11):

$$\theta_B = \arcsin \sqrt{\frac{\varepsilon}{\varepsilon + 1}} \tag{3.13}$$

By definition, vertical polarization means that the electric field vector lies in the plane of the incident angle. Similarly, horizontal polarization means that the electric field is perpendicular to that plane. This causes no problems when using R_v and R_h for horizontal or vertical surfaces, but in other cases the wave has to be divided into two components, vertical and horizontal, with regard to the surface element. Then R_v and R_h, respectively, are used for these components.

For highly conductive surfaces (e.g. sea water) and surfaces with a high dielectric constant (e.g. fresh water) the horizontal reflected field is in antiphase with the incident field (i.e. $R_h \approx -1$), independent of

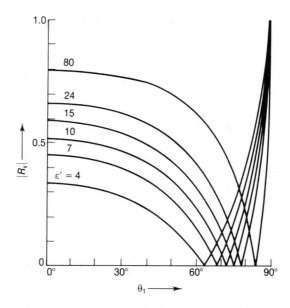

Figure 3.1 The Fresnel reflection coefficient for vertical polarization and a real dielectric constant

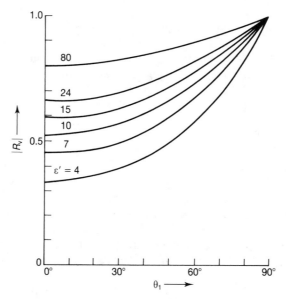

Figure 3.2 The Fresnel reflection coefficient for horizontal polarization and a real dielectric constant

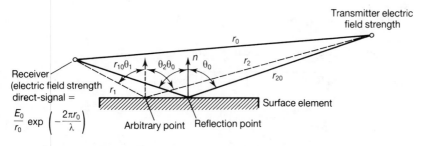

Figure 3.3 Specular reflection geometry

the angle of incidence. In these cases, the vertical reflected field is of the same order and sign as the incident field (i.e. $R_v \approx +1$), with the exception of nearly grazing angles along with high dielectric constant, in which case $R_v \approx 0$ (Equation (3.13) and Figure 3.1). Circular polarization (i.e. the field vector is rotating at the carrier frequency) can be regarded as a vertical and a horizontal field vector in phase quadrature. Because of the behaviour of the horizontal component at reflection, the direction of rotation is reversed, i.e. right-hand circular polarization is turned into left-hand and vice versa. This phenomenon may be utilized for discriminating against reflected waves.

The output signal from the receiving antenna usually contains three principal components: direct signal, specular reflections and diffuse reflections. These components add up vectorially at the antenna and are weighted by the antenna diagram according to frequency and direction of incidence. The direct signal is

$$V_d(t, \theta, \Phi) = G(\theta, \Phi)A(t)\cos[\Omega t + \Psi(t)] \tag{3.14}$$

where θ and Φ are direction angles to the transmitter (e.g. in azimuth and elevation), G is the antenna gain, and A and Ψ are the amplitude and angle modulations, respectively. Each of the reflected components is characterized by a similar expression (Forssell, 1974):

$$V_r(t_k, \theta_k, \Phi_k) = G(\theta_k, \Phi_k)r_k A(t - t_k)$$
$$\times \cos[\Omega(1 + \dot{t}_k)(t - t_k) + \Psi(t - t_k)] \tag{3.15}$$

where r_k is the complex reflection coefficient of the kth reflection, θ_k and Φ_k denote the position of the reflecting surface, t_k is the delay of the kth reflected component and $\dot{t}_k$ is its time derivative.

The total signal received is

$$V(t) = V_d(t, \theta, \Phi) + \sum_{k=1}^{n} V_r(t_k, \theta_k, \Phi_k) \qquad (3.16)$$

whose amplitude and phase can be computed as

$$|V(t)| = \sqrt{[\mathrm{Re}\, V(t)]^2 + [\mathrm{Im}\, V(t)]^2} \qquad (3.17)$$

and

$$\arg[V(t)] = \arctan\left[\frac{\mathrm{Im}\, V(t)}{\mathrm{Re}\, V(t)}\right] \qquad (3.18)$$

3.3 The ground wave

As an example from everyday life, it could be mentioned that daytime reception from a standard medium-wave or long-wave transmitter is based on ground wave propagation. The waves are called ground waves because an essential part of the wave energy follows the curvature of the earth because of diffraction at the earth's surface, absorption in the earth and refraction in the atmosphere. Up to about 3 MHz these waves predominate in the daytime. The range is greater for low frequencies than for high ones as the numerical distance (see below) increases with frequency.

The propagation of ground waves depends on the conductivity and dielectric constant of the earth, in such a way that received power is inversely proportional to the distance in the fourth power instead of, as in free space, the distance in the second power (David and Voge, 1969). In addition, other losses also increase with frequency.

Based on the radiation from a vertical antenna over plane earth with limited conductivity, wave propagation can be described mathematically using Maxwell's equations and relevant boundary conditions (tangential fields should be continuous), but this leads to very complicated expressions and is not within the scope of this book. The following presentation, therefore, is limited to an overview of the practical results, which is more suitable in this case.

The field strength of the ground wave is the vector (Norton, 1937)

$$\mathbf{E}_F = K(1 - R_v)F\, \frac{\exp\left(-j\,\dfrac{2\pi}{\lambda}\, r\right)}{r}\, [(1 - u)\mathbf{e}_1 + u\sqrt{1 - u^2}\,\mathbf{e}_2] \qquad (3.19)$$

Here R_v is the reflection coefficient of Fresnel for a vertically polarized wave (Equation (3.11)), K is a constant which is proportional to the length of the transmitting antenna, its feed current and frequency, r is the distance to the transmitter antenna, and $\mathbf{e}_1$ and $\mathbf{e}_2$ are unity vectors parallel and perpendicular, respectively, to the vertical dipole which is the transmitter antenna, F is an attenuation factor which, if on the earth's surface, can be written (Norton, 1937)

$$F = 1 - j\sqrt{\pi\xi}\, e^{-\xi}\, \text{erfc}\,(j\sqrt{\xi}) \tag{3.20}$$

where

$$\xi = \frac{j\pi}{\lambda}\, ru^2(u^2 - 1) = p\exp\,(jb) = p_1 \tag{3.21}$$

$$u^2 = (\varepsilon' - j60\sigma\lambda)^{-1} \tag{3.22}$$

where σ is the conductivity of the earth in S/m, λ is the vacuum wave length in m and $\varepsilon' - j60\sigma\lambda$ is the relative dielectric constant or permittivity of the earth:

$$\text{erfc}\,(x) = \frac{2}{\sqrt{\pi}} \int_x^{\infty} \exp\,(-y^2)\, dy \tag{3.23}$$

Equation (3.19) shows very clearly what happens when the wave is propagated on the earth's surface. If the earth were a good conductor, u (according to Equation (3.22)) would be 0, and then the extra attenuation constant would be $F = 1$, and the wave would be vertically polarized since the coefficient along $\mathbf{e}_2$ (propagation direction), would be 0. This also applies approximately for sea water with $\varepsilon' = 80$ and $\sigma = 5$ S/m (roughly up to the microwave range). However, for land, maybe with $\varepsilon' = 5$ and conductivity as low as 10^{-3} S/m, losses become much greater ($|F|$ grows), and the polarization vector is inclined forward in the direction of propagation.

The attenuation factor, F, indicates the increase of the wave attenuation compared to free space propagation. As shown by Equations (3.20)–(3.23), it depends on geometrical distance as well as on conductivity and frequency, which, together, are called the numerical distance p in Equation (3.21). Figure 3.4 shows $|F|$ as a function of p, while Figures 3.5 and 3.6 show the influence of each of the factors.

The forward slope of the field vector when the wave propagates over land with low conductivity adds a component to the Pointing vector directed towards the earth's surface. This component represents that part of the power which is dissipated in the earth's surface. Usually, the two field components are not in phase, which then leads

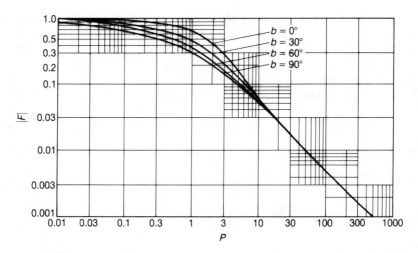

Figure 3.4 Additional attenuation of the ground wave as a function of numerical distance (Norton, 1936)

to an elliptical polarization of the field near the earth's surface.

Equation (3.19) also shows that the phase of the received signal varies with the properties of the surface. This is of importance in transmission systems where the phase contains information, e.g. in several of the most common navigation systems, and it will thus lead to different results depending on the propagation path of the received wave. For this reason DECCA, for instance, has a map of corrections taking the dependence of the phase velocity on directions from different transmitters into consideration. Figure 3.7 gives a few examples of phase corrections at 100 kHz. The diagram shows that low conductivity implies that the attenuation of the ground wave is increased considerably when the frequency increases, and also when the conductivity decreases. Table 3.1 gives examples of what this may mean (the table applies to 100 kHz, i.e. both DECCA and LORAN-C). Large areas of Norway have very bad propagation conditions where the last two columns of Table 3.1 apply. This is also the case for large areas of Iceland and Greenland.

The marked frequency dependence shown by Figures 3.5 and 3.6 is exemplified in Table 3.2.

All the formulas and figures given in Tables 3.1–3.2 and Figures 3.6–3.7 below apply to vertical polarization. However, relations are basically similar for horizontal polarization also. Even in this case the result is a certain elliptical polarization, but the ellipticity is somewhat different to that for vertical polarization.

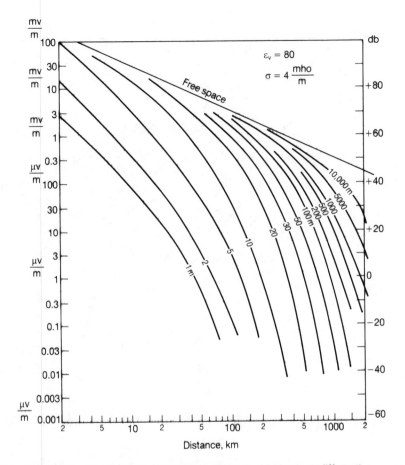

Figure 3.5 The relative field strength of the ground wave at different frequencies as a function of distance, over sea (reprinted with permission from P. David and J. Voge: *Propagation of Waves*, 1969, Pergamon Press)

The additional attenuation in the case of horizontal polarization is (Norton, 1937)

$$G = 1 - j\sqrt{\pi\zeta}\,\mathrm{erfc}\,(j\sqrt{\zeta}) \qquad (3.24)$$

where

$$\zeta = \frac{j\pi r(u^2 - 1)}{\lambda u^2} \qquad (3.25)$$

and u is given by Equation (3.22). For large numerical distances, i.e. values of the real part of ζ, G approaches u^4F. As

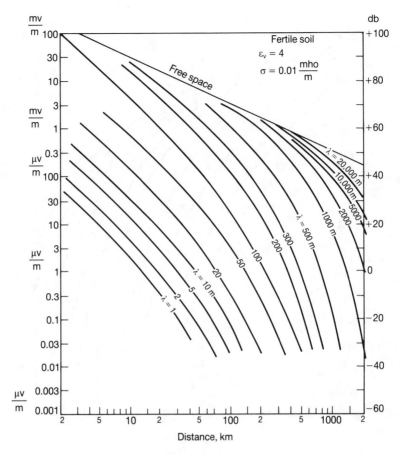

Figure 3.6 The relative field strength of the ground wave at different frequencies as a function of distance, over land (reprinted with permission from P. David and J. Voge: *Propagation of Waves*, 1969, Pergamon Press)

$u^2 = (\varepsilon - j60\sigma\lambda)^{-1}$ is always much smaller than unity, the horizontally polarized wave is attenuated much faster than a vertically polarized wave at the same frequency. The same curve of attenuation (Figure 3.4) can be utilized for both polarizations, but for a given distance, r, the numerical distance, p, is greater for horizontal than for vertical polarization. At lower frequencies where u^2 is very small (large λ, see Equation (3.22)), this difference is accentuated. Then only vertical polarization is practical, and in practice most long-range systems use vertical polarization. All antennae where one end is coupled to the earth also give vertical polarization.

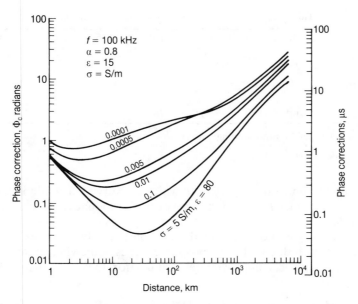

Figure 3.7 Phase correction of the ground wave at 100 kHz over sea water, (at the bottom) and at different conductivities (Gløersen *et al.*, 1974)

Table 3.1 Additional attenuation (in relation to free space)

	Free space	Sea	Good soil	Bad soil, sandy desert	Mountains generally	Other mountains, ice, dry snow
Conductivity (s/m)		5	10^{-2}	10^{-3}	5×10^{-4}	$10^{-4} - 10^{-5}$
Additional attenuation (dB)	0	9	11	24	39	63–160

Notes: distance = 1000 km; frequency = 100 kHz.

Source: Gløersen *et al.*, 1974.

3.4 Tropospheric influence

The atmosphere of the earth does not have distinct limits, but it is often divided into spherical layers according to certain meteorological criteria. The region nearest to the ground is called the *troposphere*

Table 3.2 Relative field strength over sea and land

Frequency	(MHz)	0.1	0.5	1.0	5	10	50
Relative field strength (dB)	sea	66	65	64	60	56	18
	land	63	34	28	−9	−20	−50

Notes: distance 150 km; transmitting antenna, a vertical whip.

Source: Gløersen et al., 1974.

and stretches up to the *tropopause* which in Norway lies at an altitude of about 10 km. The region above the tropopause is called the *stratosphere*.

Because of gas molecules in the air, particularly of water vapour, the refraction index of the troposphere, n, is somewhat greater than unity. An empirical approximate formula for the refractivity, N, is (Kerr, 1951)

$$N = (n - 1)10^6 = \frac{77.6}{T}\left(p + \frac{4810e}{T}\right) \tag{3.26}$$

where T is the temperature in Kelvin, p is the total pressure and e is the partial pressure of the water vapour, both in mbar. However, N is a function of altitude (because all parameters of Equation (3.26) vary with the altitude). This function could be described approximately as a linear decrease from sea level up 1–2 km, an exponential decrease from there to the tropopause (where constant temperature but powerful winds dominate) and N is of the order of 100 (compared to about 300 on the earth's surface), and a new exponential decrease as a function of altitude in the stratosphere. An approximation of tropospheric conditions is (Figure 3.8)

$$N = N_s \exp(-qh) \tag{3.27}$$

where N_s is the value on the earth's surface and h is the altitude. Conventionally, a so-called standard atmosphere is used in calculations, i.e. $q = 0.136 \times 10^{-3}(\text{m}^{-1})$.

Refractivity N can be expanded in a series as a function of altitude (for $h \lesssim 8$ km), which shows the linearity next to the earth's surface:

$$N \approx N_s - \frac{h}{4a} 10^6 = N_s - 0.039h \tag{3.28}$$

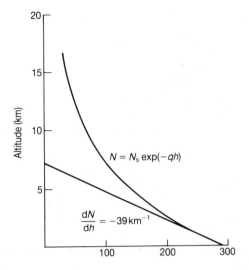

Figure 3.8 Refractivity variation as a function of altitude in a standard atmosphere ($N_s = 290$)

where a is the earth radius.

There are usually great deviations from the standard atmosphere because of regular diurnal and seasonal variations as well as faster irregular fluctuations.

The reason for the interest in refraction index variations in a navigational context is that the propagation velocity of the signals is influenced. In free space we have

$$\frac{v}{v_0} = \frac{\lambda}{\lambda_0} = \frac{1}{\sqrt{\varepsilon'}} = \frac{1}{n} \tag{3.29}$$

where v and v_0 are velocities in the medium and vacuum, respectively, λ and λ_0 are the corresponding wavelengths, and ε' is the relative dielectric constant (the real part) of the medium. As time is the real measurement quantity of all navigation systems based on measurements of distance or distance differences, the propagation velocity of the signals must be known if the transformation into distance is not to lead to errors. Because of Snell's refraction law, variations in n also imply that the signals in the air do not propagate along a straight line (Figure 3.9), but this effect can usually be dismissed in a navigational context. Snell's law tells us that

$$n_1 \sin \Psi_1 = n_2 \sin \Psi_2 \tag{3.30}$$

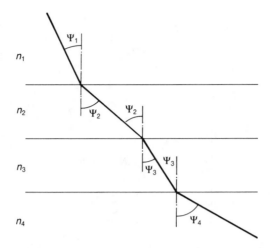

Figure 3.9 A plane layered medium with refraction index n_i

i.e. passing from one medium with a lower refraction index the ray is refracted away from the perpendicular. (The derivation of this law can be performed simply by using two plane waves at each side of the boundary, and is left to the reader.) Among other things, this causes a signal in the air transmitted parallel to the earth's surface to follow a path which curves in the same direction as the earth. It can be shown that with a standard atmosphere (Equation (3.28)) the signal path would hold a constant altitude if the earth's radius were $\frac{4}{3}$ of the real one, or in other words, the radio horizon is at a distance equal to the geometrical horizon on an earth having a radius $\frac{4}{3}$ times the real one.

The propagation velocity of the ground wave is also influenced by the refraction index of the troposphere, although the influence is small in relation to that of the conductivity of the earth.

The reason for this influence is that part of the signal energy is propagated in the air, for even if the wave follows the earth's surface there are corresponding electric fields above, as well as below, the ground plane. These fields decrease exponentially (in different ways) with distance to the ground plane. Therefore, there may be a measurable influence on the signal propagation velocity by the conductivity at several hundred metres' depth and by the refraction index at several kilometres' altitude.

All the navigation systems presented in the following chapters contain computational subsystems where different assumptions are made about the propagation velocity of the signals; and deviations of course lead to position errors. Such deviations, are, for example, weather changes, particularly so-called 'fronts' which can change pressure, temperature and humidity fairly rapidly. In LORAN-C and DECCA, for example, the passage of such fronts may lead to considerable position errors.

The refraction index is also a little influenced by precipitation. The influence is small but it increases with frequency. Below about 10 GHz and at moderate precipitation intensity it is usually not necessary to consider such influence.

Satellite navigation receivers sometimes have facilities for corrections based on atmospheric data, Equation (3.26) and a model of the altitude variation of the refraction including the elevation angle to the satellite.

3.5 Ionospheric influence

The ionosphere contains different free, neutral and charged particles, mainly electrons and ions. Thus it is a so-called *plasma*, created and sustained by solar radiation. Solar radiation is, in itself, partly a current of charged particles (the solar wind), partly ultraviolet radiation or soft X-ray radiation ionizing neutral gas molecules. The ionizing part of the solar radiation only includes wavelengths below 0.125 μm (the wavelength of visible light is 0.4–0.8 μm). The power density at these wavelengths is approximately 0.01 W/m^2, compared with the total solar radiation of about 1350 W/m^2. However, the short-wave solar radiation influencing the ionosphere depends much more strongly on sun spot activity than on total radiation. The properties of the ionosphere as a function of altitude, therefore, vary not only with the time of day and year, but also with sun spot activity and sudden eruptions from the sun. The two most important ionospheric effects caused by the latter are the so-called SID (Sudden Ionospheric Disturbance), and PCA (Polar Cap Absorption). SID is caused by X-ray eruptions from the sun resulting in a lower ionosphere on the sunny side of the earth; 100–200 SIDs a year are common, and the duration of the disturbance usually varies from a few minutes to a few hours each time. PCA is caused by proton bursts from the sun which are deflected towards the magnetic poles due to the magnetic field of the earth. These disturbances can have a duration of several days.

Knowledge of the ionosphere and of its influence on radio waves is relatively new, and the ionosphere continues to be an object of fruitful research. The term 'ionosphere' was proposed by the British physicist Watson-Watt in 1926. The denominations D, E and F of the different regions of the ionosphere (Figures 3.10, 3.11 and 3.12) were proposed by another British physicist, Appleton.

The increasing use of satellite systems for precise position measurements has provoked an increased interest in ionospheric behaviour, particularly propagation properties as functions of time and position.

Figure 3.12 shows that the different ionospheric regions are in truth not as sharply defined from each other as radar echoes, ionograms, etc., may indicate. (In ionograms the altitude of the reflecting region is measured by means of radar pulses, and an effective reflection altitude is measured based on the time of arrival of the point of gravity of the reflected radar pulse.) The diagram also shows that the concentration of neutral particles, n_g, falls rapidly with increasing altitude.

Collisions between electrons and neutral particles give some wave absorption in the E-region and strong absorption in the D-region and at lower altitudes. Absorption where reflection takes place is particu-

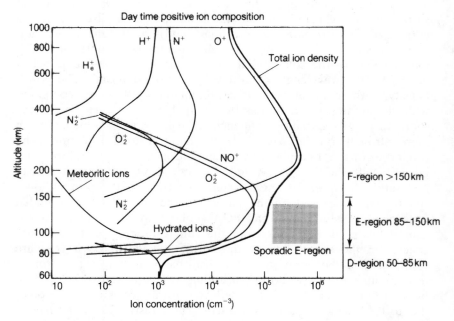

Figure 3.10 Ion profiles in the middle of the day at mid-latitudes and sun spot minimum (Corrigan and Skrivanck, 1974)

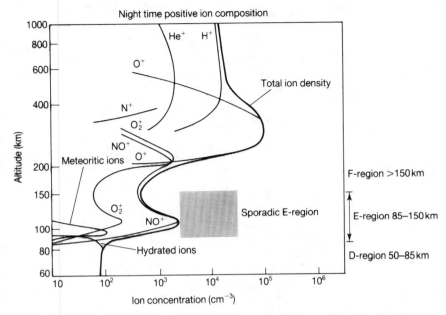

Figure 3.11 Ion profiles in the middle of the night at mid-latitudes and sun spot minimum (Corrigan and Skrivanck, 1974)

larly strong because the wave then has a small group velocity and thus stays in an absorptive environment for a relatively long period.

The wave reflection is most simply explained optically by means of the refractive index. An electromagnetic wave in a plasma influences the movement of the charged particles by means of its field strength. It causes collisions between the particles, but as these are charged they also, by their movements, generate a field having the same character as the incident wave field. The electromagnetic wave thus moves in a kind of dielectric. The relative dielectric constant of a plasma with no magnetic field turns out to be (Kelso, 1964)

$$K = 1 - \frac{(\omega_p/\omega)^2}{1 + (v_c/\omega)^2} - j\,\frac{v_c}{\omega}\,\frac{(\omega_p/\omega)^2}{1 + (v_c/\omega)^2} \tag{3.31}$$

where ω is the angular frequency of the wave, v_c the collision frequency and ω_p the plasma frequency. The latter quantity can be derived superficially in the following way, if $v_c \ll \omega$. (v_c is in the MHz range in the lowest parts of the ionosphere and decreases exponentially with height.)

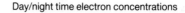

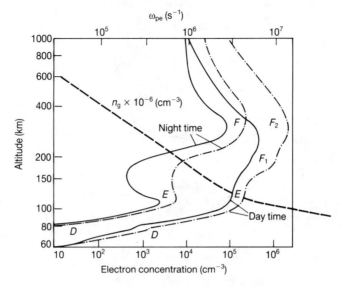

Figure 3.12 Typical electron density profiles in the middle of the day and in the middle of the night at mid-latitudes and sun spot minimum (solid line) or sun spot maximum (stippled line) (Corrigan and Skrivanck, 1974)

An electron having the mass m_e and charge q_e is influenced by an electric field $E = E_0 \exp(j\omega_p t)$. The electron then develops the acceleration a and the velocity v because of the force F, where

$$F = m_e a = m_e \frac{dv}{dt} = jm_e \omega_p v = -q_e E \tag{3.32}$$

The magnetic field of the earth is static and, according to Maxwell, we then have

$$\frac{\partial D}{\partial t} + i = 0 \tag{3.33}$$

where $\partial D / \partial t$ is the displacement current and

$$i = -N_e q_e v \tag{3.34}$$

is the current density because of the electron movement (N_e is the number of electrons/unit volume). Equations (3.33) and (3.34) inserted into Equation (3.32) then give

$$-N_e q_e v = -\frac{\partial D}{\partial t} = -j\omega_p \varepsilon_0 E = -j\omega_p \varepsilon_0 \frac{j\omega_p m_e v}{q_e} \tag{3.35}$$

Equation (3.35) then states that

$$\omega_p^2 = \frac{N_e q_e^2}{\varepsilon_0 m_e} \tag{3.36}$$

$f_p (= \omega_p/2\pi)$ is thus a cut-off frequency which can be taken as the highest frequency of a signal whose oscillations the electrons are able to follow. Higher signal frequencies do not give corresponding movements to the electrons because of their masses. (As ions are at least about 1800 times heavier than electrons, they have considerably lower cut-off frequencies, even if the same phenomenon is, in principle, also valid for them.) The imaginary part of K determines the losses in the plasma, while the real part determines the phase velocity. Then $\text{Re}(K) = 1 - (\omega_p/\omega)^2$. The refraction index is

$$n = \sqrt{\text{Re}(K)} = \sqrt{1 - \left(\frac{\omega_p}{\omega}\right)^2} \tag{3.37}$$

The propagation velocity of the electromagnetic wave in the plasma is

$$v_f = c/n \tag{3.38}$$

where c is the velocity in a vacuum. Equation (3.38) shows that v_f is imaginary when $\omega < \omega_p$, which implies that the wave can propagate in the plasma only for frequencies above the plasma frequency. Here we have the explanation of the ionospheric reflection: signals at higher frequencies pass through the ionosphere, while those at lower frequencies are reflected and attenuated. A reflection coefficient can be expressed as a function of the dielectric constant of the plasma K, the incident angle and the polarization in the same way as for other reflection surfaces by means of the reflection coefficients of Fresnel.

That part of the power which is not reflected passes into the plasma (the ionosphere), but changes its direction in passage according to Snell's law of refraction (Figure 3.9 and Equation (3.30)).

As the refraction index of the ionosphere depends on the varying electron and ion concentrations (Equations (3.36) and (3.37)), although distinct boundaries do not exist, reflection and refraction take place in a more or less layered medium (Figure 3.13). As the

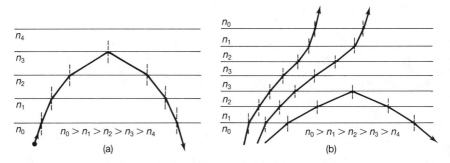

Figure 3.13 Refraction in a layered medium

phenomenon at the boundary layers is frequency- and angle-dependent, the ray path is similar to that shown in Figure 3.14.

Thus signals at a low frequency are reflected at low altitudes, while high-frequency signals are only refracted, but they may develop an increased angle of incidence so that they are reflected at a higher layer. Roughly, frequencies below about 500 kHz (among them DECCA and LORAN signals) are reflected in the D-region, 0.5–2 MHz in the E-region and 5–30 MHz in the F-region, whereas frequencies above 30 MHz are not reflected in the ionosphere. However, large variations may occur. As the angle of incidence has to be above a certain magnitude before reflection takes place, there exists a minimum distance on the ground between the transmitted and received sky wave, the so-called 'skip distance' (Figure 3.14). Together with the altitude of the ionosphere this gives a certain minimum time delay of the sky wave. This phenomenon is utilized in LORAN-C where detection of the arrived pulse is assumed to take place before the arrival of the sky wave (Section 4.5).

It can also be deduced from the preceding paragraph and Figure 3.14 that signals using frequencies higher than a certain frequency are not reflected by the ionosphere. This is called the maxim usable frequency (MUF) in communication terminology.

It should be mentioned that in the presence of a magnetic field, e.g. the earth's magnetic field, Equation (3.31) which is an approximation of Appleton–Hartree's formula, could be written as:

$$K = 1 - \frac{\omega_\mathrm{p}^2}{\omega(\omega \pm \omega_\mathrm{H} \cos \theta - j v_\mathrm{c})} \tag{3.39}$$

where ω_H is the gyrofrequency (usually 1–2 MHz in the ionosphere)

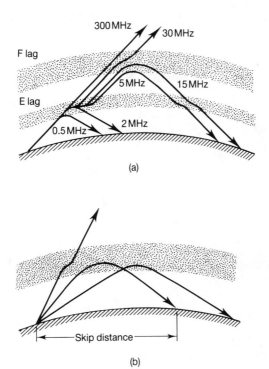

(a)

(b)

Figure 3.14 Refraction and reflection in the ionosphere: (**a**) different; (**b**) equal frequencies

$$\omega_H = \frac{-q_e B_e}{m_e} \tag{3.40}$$

θ is the angle between the direction of propagation and the magnetic field, and B_e is the flux density of the magnetic field. The plus sign is for the ordinary wave, and the minus sign for the extraordinary wave.

System descriptions

4 Hyperbolic systems

4.1 Introduction

The following description of hyperbolic systems is preceded by some generalities common to all hyperbolic systems. The description itself deals with the most common long-range/medium-range systems, namely OMEGA, DECCA and LORAN-C. The abundance of short-range hyperbolic systems (e.g. 2 MHz and microwave systems) is not touched upon, mainly to keep the presentation within reasonable limits.

4.2 General

The geometrical influence on the performance of navigation and positioning systems in the hyperbolic mode has been treated in Chapter 2. There, it is shown that hyperbolic systems and $\rho-\rho$ systems are very similar with regard to the geometrical influence. For example, the displacement of the lines of position as a function of the measurement error is, in both cases, inversely proportional to the sine of half the aspect angle to the transmitters. The main difference is that it is sufficient to have two transmitting stations in $\rho-\rho$ navigation, whereas the hyperbolic mode requires three.

For a discussion of $\rho-\rho$ versus hyperbolic systems, let us assume that the system has an unlimited number of users. Then, only the user equipment should receive, and transponders are ruled out. The transmitters at the reference stations must have a common accurate time reference so that the times of transmission can be related to each other with small and known errors. The receiver, on the other hand, usually has a much less accurate time reference and thus has difficulty in relating the received signals to the time reference of the transmitters. This gives rise to a time error and hence also to a position error. A solution to this problem (utilized in today's satellite receivers) is to use the deviation between system time and receiver time (or frequency) as an additional unknown in the system of equations for calculation of position. With the calculation capacity (and oscillator stability) of the receivers of the 1950s and 1960s, however, this was not possible. In such a situation a hyperbolic system is the solution. Then, the received signals from two transmitters are utilized for measurement of time or phase differences, and all errors common to the two signals are eliminated, among them the clock error of the receiver.

Based on the computational capacity of today's receivers all terrestrial hyperbolic systems should be able to use the $\rho-\rho$ mode with time as the third unknown, assuming that the time of transmission is known (to the receiver) by means of some signal. In certain situations this may give a geometrically better solution as the LOPs are reconfigured.

Increased computational capacity of the receivers may also be utilized to refine the computations because part of the clock deviation between the transmitter and the receiver is known. After a few position determinations in $\rho-\rho$ plus time mode, a filtered clock deviation may be computed, and in a system of equations this can be regarded as a known parameter, whereas drift and noise are the unknown quantities as far as the clock is concerned. Thus, the clock deviation is modelled as a time-dependent function where the constant term is known and only first-order (and possibly second-order) terms have to be computed. This will increase the accuracy of the subsequent position determinations further (Hatch, 1978).

4.3 Phase measurements

In hyperbolic navigation systems, phase measurements are made regularly between signals from different transmitters within one system. Phase measurement between different navigation systems (e.g. one OMEGA and one LORAN transmission) are comparatively rare.

The simple principle is shown in Figure 4.1. Two equal-frequency

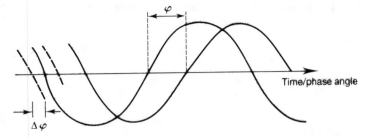

Figure 4.1 Principle of the measurement of phase differences with noise influence

signals with the phase difference φ are compared.

We assume that the sinusoidal signal has the power S and thus the peak amplitude $\sqrt{2S}$. The noise power in the bandwidth used is assumed to be N, i.e. $\overline{[n(t)]^2} = N$, where $n(t)$ is the noise voltage. Thus the signal-to-noise ratio is S/N. The signal voltage then has the shape $\sqrt{2S} \sin \varphi$, where $\varphi = \omega t$ and ω is the angular frequency. The noise voltage corresponds to a phase displacement $\Delta\varphi$ of the noise-free signal

$$\sqrt{2S} \cos \varphi \Delta\varphi = n(t) \tag{4.1}$$

Squaring and taking the time average on both sides of Equation (4.1) we get

$$\sqrt{(\Delta\varphi)^2} = 1 / \sqrt{\frac{S}{N}} \tag{4.2}$$

In measurements of phase differences between two signals both containing uncorrelated noise, the errors add quadratically, and the resulting rms error is

$$\Delta\varphi_{1,2} = \sqrt{\frac{1}{(S/N)_1} + \frac{1}{(S/N)_2}} \tag{4.3}$$

(It should be added that signals from different transmitters in a hyperbolic system may contain correlated noise.)

The same method as that given above can be utilized to evaluate time errors at the detection of pulsed signals. The pulse width, τ, is often measured between the $-3\,\text{dB}$ points, which, on a sine curve, corresponds to the angles of $\pi/4$ and $3\pi/4$. With the detection error of Δt we then have, approximately,

$$\frac{\Delta t}{\tau} = \frac{\Delta \varphi_\mathrm{p}}{\pi/2} \tag{4.4}$$

i.e.

$$\Delta t = \frac{2\tau}{\pi \sqrt{S/N}} \tag{4.5}$$

Any phase measurement has an ambiguity of $n \times 2\pi$, implying that a certain phase difference describes an unlimited number of hyperbolas with a mutual separation corresponding to a change of the phase difference by 2π. Far away from the transmitters, such a change would be a special case of the displacement of the lines of position treated in detail in Section 2.2.1.2. This means that the distance between these hyperbolas is inversely proportional to the sine of half the aspect angle, i.e.

$$\Delta d \sim 1/\sin\frac{\alpha}{2} \tag{4.6}$$

In order to find the constant of proportionality it is sufficient to study what happens on the baseline, where $\alpha = 180°$. If the receiver is moved half a wavelength towards one of the transmitters, the phase of the received signal from this transmitter will decrease by π, but at the same time the phase of the signal from the other transmitter will increase by the same amount. Thus the difference changes by 2π. The widths of the non-ambiguity area can thus be written

$$\Delta d = \frac{\lambda/2}{\sin \alpha/2} \tag{4.7}$$

where α is the aspect angle to the transmitters. The strip between the two hyperbolas with a mutual separation of Δd is called a lane. The lane width is characteristic of the system because it is directly dependent on signal frequency.

The phase measurement itself can, for example, be realized by letting the rising edge zero-crossing of one of the signals trigger a counter, and the equivalent zero-crossing of the other signal stop the counter. The use of hard limiting and zero-crossings is preferable in the sense that this reduces the influence by differences in signal levels. On the other hand, hard limiting reduces the signal-to-noise ratio a little.

4.4 OMEGA

4.4.1 General

OMEGA is a hyperbolic navigation system based on phase difference measurements, operating at 10–14 kHz. The deployment of the system began in the mid-1960s after a period of test runs on a few transmitters, but the history of the system dates back to the years immediately after World War II. Before the introduction of OMEGA, a considerable amount of research and experiment on the use of VLF signals for phase comparison systems had been carried out. The interested reader is referred to the literature (e.g. Pierce, 1965, 1989) for historical details.

The advantages of the system stem from the use of the very low frequencies enabling the system to cover the whole of the earth's surface by eight transmitting stations. The stations are listed in Table 4.1.

Like many other navigation systems OMEGA was originally intended for military use, but the number of civil users has gradually increased. Today (Spring 1990) it is the only navigation system with time-continuous, global coverage.

4.4.2 Signal format

Initially, the OMEGA system operated only at frequencies 10.2, $11\frac{1}{3}$ and 13.6 kHz. The first frequency is the main frequency and the one that all receivers can receive. The other two have been used mainly to enlarge the non-ambiguity area. (The resolution of ambiguities is explained in Section 4.4.4.) The lane width (Equation (4.7)) at 10.2 kHz is about 15 km, but at the difference frequency ($11\frac{1}{3}$–10.2 kHz) it is about 132 km. However, it became necessary to enlarge the non-ambiguity area further, and a fourth transmitter frequency, 11.05 kHz, was introduced. The difference frequency between this one and $11\frac{1}{3}$ kHz gives a lane width of about 529 km. In addition, a frequency of identification was introduced for each transmitting station. These changes, which were brought about in 1976–9 give a final signal format, as shown in Figure 4.2. Frequencies

Table 4.1 OMEGA transmitter stations

Station	Location	Transmitter antenna	Administered by
A	Bratland, Norway	Suspended wires across a fjord	Norwegian Telecommunications Administration
B	Monrovia, Liberia	Grounded tower with radial top elements	Ministry of Industry and Commerce
C	Haiku, Hawaii	as A	US Coast Guard
D	La Moure, North Dakota	Bottom-isolated monopole	As C
E	Reunion in the Indian Ocean (France)	as B	French Navy
F	Golfo Nuevo, Argentina	as D	Argentine Navy
G	Woodside, Victoria, Australia	as B	Department of Transport
H	Tsushima, Korean Strait, Japan	as D	Japanese Coast Guard

F_1, which are unique to each station, are 12.1, 12.0, 11.55, 13.1, 12.3, 12.9, 13.0 and 12.8 kHz, for transmitters A to H respectively. These signals are also used for calibration purposes.

As shown by Figure 4.2, there is a time interval of 0.2 s between each segment of the signal format, and the repetition frequency of the signals is 0.1 Hz. The lengths of the signal segments are 0.9, 1.0, 1.1 or 1.2 s. The main reason for using the 0.2 s spaces between transmissions is that the transit time of the signals around the earth is about 0.13 s, and, consequently, the signals of each segment have time to decay before the start of the next segment. Another reason is that the transmitter antenna circuit has to be retuned before a frequency shift.

The signal sequence starts at every tenth second of UTC (Universal Time Coordinated). All transmitters are controlled by cesium clocks, and there is no phase locking between them. Consequently, no master–slave relation exists in DECCA terminology (Section 4.5.1). System time is determined with the reference source of the US Naval

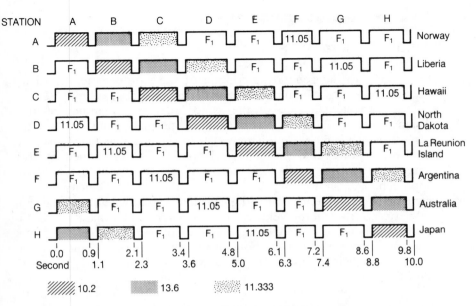

Figure 4.2 Signal format of the OMEGA system

Observatory. Individual clock errors at the transmitting stations are kept within 2 μs rms and are not a dominating source of error in the system. Transmitter corrections are brought about with the aid of satellite transferred signals.

The Soviet Union has its own navigation system at VLF with transmitters at Krasnodar, Novosibirsk and Komsomolskamur. Unfortunately, this system is not compatible with OMEGA as the frequencies are 11.905, 12.649 and 14.881 kHz. The repetition time of the signal sequence is 3.6 s, making the system more suitable for aircraft navigation than OMEGA. Radiated power is 50–100 kW (Beukers, 1974).

Several communication transmitters also operate at VLF, most of them in the frequency range 15–20 kHz, but also a few in the range 20–25 kHz. These transmitters are very phase stable, transmitting continuously with high power, 50–1000 kW. Their signals can also be utilized for navigation, but as this is not the main purpose, and in the event that the transmitters are switched off, they cannot be utilized alone. In addition correction data are not published to the same extent for these transmitters as for OMEGA (Section 4.4.11). Some professional quality OMEGA receivers are able to use these transmissions for navigation together with OMEGA signals.

4.4.3 Wave propagation at VLF

4.4.3.1 General

In addition to the general description of wave propagation given in Chapter 3, we describe the particular phenomena appearing at VLF.

These wavelengths are of the same order of magnitude as the distance between the D-region of the ionosphere and the earth's surface, which is 50–90 km. The wave propagation thus takes place in a spherical waveguide whose (lossy) boundaries are formed by the ionosphere and the earth respectively. Figure 4.3 shows a simplified approach. As shown by the diagram, the phase velocity, $f\lambda_f = c/\cos\alpha$, is greater than that of light in free space, whereas the group velocity, i.e. the energy velocity, $f\lambda_g = c\cos\alpha$, is less than this. Both of them depend on the angle α, i.e. the ionospheric altitude, meaning that they vary with the conditions of the ionosphere, e.g. the time of the day and year (Figure 4.4). The curves of Figure 4.5 show that the variations depending on the increase and decrease of the effective height of the waveguide during the day may introduce large errors (several km), even in a system measuring differential phase. The reason for this is the difference of the influence of the sun at different times at the two transmitting stations.

The effective height generally varies between 60 and 90 km. Most of this variation can be predicted so that maps of corrections as functions of time and place can be worked out. The conditions are complicated because there exist several waveguide modes, i.e. partial waves with different α and hence different phase velocity (Figure 4.3). The most important here are TM modes (Figure 4.4).

The first mode, i.e. the mode with the lowest cut-off frequency, is

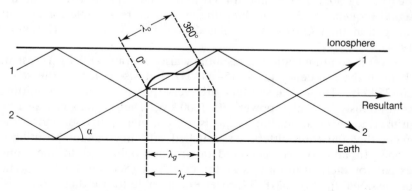

Figure 4.3 Idealized sketch of the wave propagation at VLF

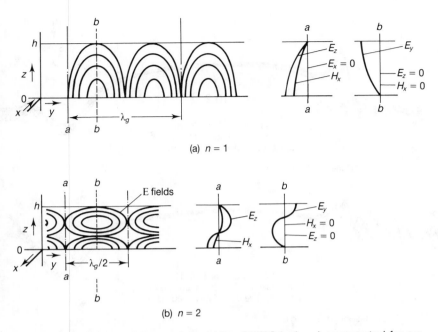

(a) $n = 1$

(b) $n = 2$

Figure 4.4 Idealized field strengths of the OMEGA signals, computed for an earth with infinite conductivity. (The height of the waveguide propagating that particular mode is $(n - 1/2)$ wavelengths.) $n = 1$ is the lowest (and useful) mode

used for navigation because it is less attenuated than higher modes over long distances, and because its propagation parameters are less unstable under the influence of ionospheric variations.

In practice, only the second mode need be considered as all the higher modes decay very rapidly with distance to the transmitter.

The modal interference (Figures 4.6 and 4.7) is most important in the neighbourhood of the transmitter because the higher modes are attenuated faster (Figure 4.8). Therefore, such disturbances may be avoided to a great extent by utilizing signals from more distant transmitters. It is recommended to use signals from transmitters that are more than 600 n.m. (nautical miles) away (Gupta and Morris, 1986).

The modes have different propagation characteristics, and the theoretical attenuations of the first two are shown in Figure 4.8(a) and (b). Obviously, the higher modes are attenuated faster. The cut-off frequency of the first mode can be calculated from an assumed ceiling of the waveguide at 70–90 km and is then about 4 kHz.

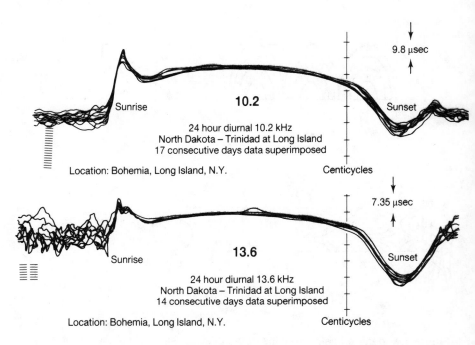

Figure 4.5 Phase differences between OMEGA signals from the stations in North Dakota and on Trinidad (closed down after the Australian station became operational), measured over 24-hour intervals in New York at (**a**) low solar activity

Practical measurements have shown a reasonable consistency with Figure 4.8(a) and (b). Among other things, it turned out that the attenuation is somewhat smaller for OMEGA signals at 13.6 kHz than at 10.2 kHz, which indicates that the tendency of the characteristic attenuation curves of a waveguide in Figure 4.8 are correct.

In order to define the coverage of an OMEGA transmitter, one often chooses the area where $S/N > -20$ dB in 100 Hz bandwidth (as this usually gives acceptable receiver performance), and, at the same time, the phase error caused by modal interference is less than 72° (20 centicycles, Figure 4.7). (20 cec is the approximate threshold above which 360° phase jumps may occur.)

Waveguide changes are caused by several different types of solar activity (see the difference between (a) and (b) in Figure 4.5) which can be predicted only to some extent and which, therefore, contribute to an increased uncertainty in phase velocity. The two most important types of solar activity erruptions are SID and PCA. The former

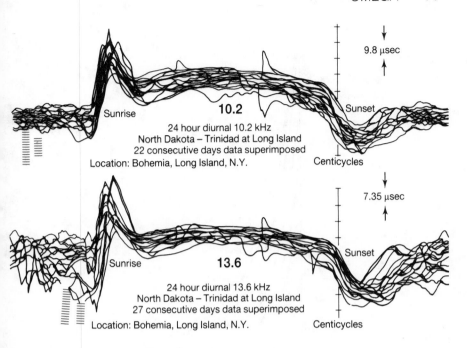

24 hour diurnal 10.2 kHz
North Dakota – Trinidad at Long Island
22 consecutive days data superimposed
Location: Bohemia, Long Island, N.Y.

24 hour diurnal 13.6 kHz
North Dakota – Trinidad at Long Island
27 consecutive days data superimposed
Location: Bohemia, Long Island, N.Y.

(b) high solar activity (Beukers, 1974)

disturbance is caused by X-rays from the sun leading to a considerable lowering of the ceiling of the waveguide in the day time, and the
latter is caused by the solar emission of proton bursts deflected
towards the magnetic poles of the earth, implying a reduced height of
the waveguide, mainly at high latitudes. Without corrections for these
anomalies the OMEGA system may give position errors up to about
10 km, in special cases up to 30 km according to measurements.

From the theory of waveguides it is known that a narrowing of a
waveguide means increased phase velocity (see also Figure 4.3). This
also applies in the propagation of VLF signals. Thus, signals propagating from west to east, i.e. from night to day (it is assumed that the
measurement takes place in daytime) have gradually increasing phase
velocity.

As the ionosphere forms only one boundary of the spherical
waveguide, the earth, which forms the other, also influences
OMEGA signal propagation. Variations in the conductivity and
permittivity of the earth thus cause corresponding propagation parameter variations.

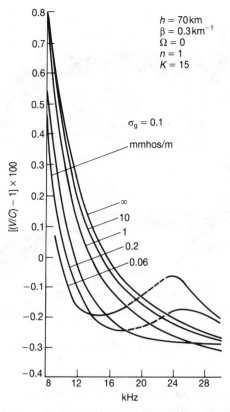

Figure 4.6 Computed phase velocity for different conductivity values as a function of frequency. Efficient ionospheric height 70 km (Wait and Spies, 1964)

The skin depths of a medium, i.e. the distance from the surface to the depth where the field strength is reduced to the surface value divided by e, is

$$\delta = \sqrt{\frac{2}{\mu\sigma\omega}} \tag{4.8}$$

where μ, σ and ω are permeability, conductivity and angular frequency respectively. The numerical values $\mu = \mu_0 = 4\pi \times 10^{-7}$ H/m, $\sigma = 5$ S/m (sea water) $- 10^{-4}$ S/m (very poor earth or mountain, dry ice) give skin depths of 3–700 m for OMEGA signals. On average, the effective conductivity equals the average between 50 and 100 m

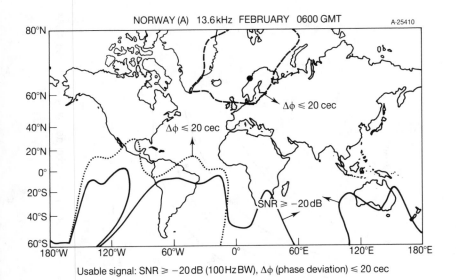

NORWAY (A)   13.6 kHz   FEBRUARY   0600 GMT   A-25410

Usable signal: SNR ≥ −20 dB (100 Hz BW), Δφ (phase deviation) ≤ 20 cec

Figure 4.7 Coverage diagram of the Norwegian station at 13.6 kHz, measured at 0600 GMT in February 1983 (Gupta and Morris, 1986)

below the surface. (In the sea the signal strength is sufficient for reception down to about 15 m.) A varying conductivity, however, implies not only varying skin depth, but also varying attenuation and phase velocity. A theoretical calculation of the latter gives curves as shown by Figure 4.6.

4.4.3.2 Conditions in the lower part of the ionosphere

The propagation of electromagnetic waves at VLF (3–30 kHz) over large distances is made possible by the large reflection coefficients in the lower ionosphere at grazing incidence. These coefficients are due to the relatively rapid altitude variation shown by the electron density in the D-region of the ionosphere (Figure 3.12). For many purposes a good approximation of the actual conditions can be achieved by regarding the ionosphere as a distinctly limited medium. However, there are cases where such a model gives erroneous results, and experiments indicate that the electron density varies as an exponential function of the altitude.

Most measurements of the properties of the ionosphere have been carried out by means of radar. Radar measurements have shown that

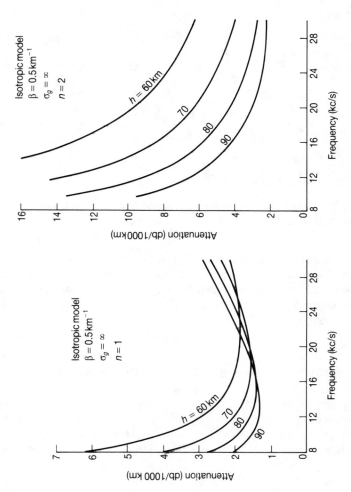

Figure 4.8 Theoretical attenuation of the 1st (**a**) and 2nd (**b**) mode. *h* is the efficient ionospheric altitude (Wait and Spies, 1964)

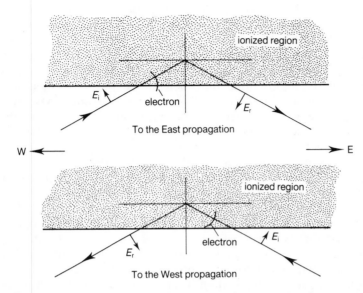

Figure 4.9 Electron movement in the lower part of the ionosphere caused by eastward and westward wave propagation, seen from the south (reprinted with permission from A. D. Watt: *VLF Radio Engineering*, 1967, Pergamon Press)

in day time there are two rather distinctly separated layers in the lower part of the ionosphere. The electron density rises fairly rapidly up to 60–70 km altitude, and reaches values of about 100 electrons per cm^3. From that altitude it is almost constant up to altitudes between 72 and 76 km and then rises again up to the bottom of the E-region, at about 85 km altitude (Figure 3.12).

4.4.3.3 The directional dependence of the reflection coefficient of the ionosphere

In the preceding sections, only the dielectric and conductivity properties of the ionosphere have been considered. However, accurate calculations of ionospheric influence of VLF signal propagation also require consideration of the earth's magnetic field.

The direct cause of ionospheric behaviour as a dielectric to an electromagnetic wave is that the electrons of the ionosphere are made to move by the wave's electric field. This movement is also influenced by the DC magnetic field of the earth. Any charged particle in a magnetic field is influenced by a force perpendicular both to the velocity vector and to the magnetic field. A vertically polarized wave such as the OMEGA signals propagating parallel to the magnetic equator, i.e. perpendicularly to the magnetic field of the earth, generates electron paths approximately as shown in Figure 4.9. Because of the almost total reflection of the wave, the electric field vector turns almost 180° (according to some investigations, at least 160°), and the velocity vector of the electrons thus forms a smaller angle to the E-field of the reflected wave (E_r in Figure 4.9) when the wave propagates eastwards, than in the case of propagation to the west. The coupling between the electric field of the reflected signal and the electron current becomes stronger for waves to the east than for waves to the west (cf. the phenomena in a microwave tube); thus the reflected signal is stronger for the easterly wave than for the westerly one, i.e. a VLF signal shows greater losses when propagating to the west than when propagating to the east. If the signal propagates along the earth's magnetic field, on the other hand, the E-field of the reflected signal forms approximately the same angle to the electron paths when propagating in the north–south and south–north directions, so there is no particular loss difference between south and north propagating waves. It should also be noted that difference between easterly and westerly propagation is most noticeable in the neighbourhood of the magnetic equator since the magnetic field in this region is parallel to the earth's surface.

The difference in attenuation between easterly and westerly propagating OMEGA signals varies with the conditions of the ionosphere, but it can be fairly large (Watt, 1967). The reflection coefficient of the ionosphere can be 0.8–0.95 for easterly waves, but 0.35–0.65 for westerly waves. The phase shift at the reflection of the wave towards the ionosphere is also closer to 180° for easterly than for westerly waves, the difference being of the order of 5–10°.

The considerable difference in gradients of attenuation between easterly and westerly propagation may imply that the signal with the shortest path from the transmitter to a receiver in the neighbourhood of the equator is weaker than that which has propagated in the opposite direction around the earth. When the two waves are about equally strong, they also interfere with each other in a standing-wave pattern where the distance between the peaks is $\lambda/2$.

In addition to the variations of the reflection coefficient mentioned above, there are variations caused by time-variable conditions in the ionosphere. The altitude of the ceiling of the waveguide has a diurnal rhythm, which influences phase velocity as well as attenuation. The phase velocity increases when the waveguide is compressed in the daytime because of increased radiation from the sun. At the same time, however, the attenuation of the signals also increases. In addition, PCA and SID and other irregular variations of the properties of the ionosphere are of some importance.

4.4.3.4 Influence of the earth and the solution of the wave equations

Because of the large wavelengths in relation to the dimensions of the transmitting antennae, these can be regarded as dipoles generating vertically polarized signals. These signals propagate in a spherical shell which has different surface impedances at the top and bottom. Calculation of the propagation constants can be made based on the conductivity and dielectric constant of the boundary surfaces and dimensions of the shell, but such calculations are very complex and are not included in this presentation. In addition to the discussion about the ionosphere sections, we consider a few supplementary aspects of the influence of the earth and how this influence can be treated.

In calculations of signal propagation, Maxwell's equations are utilized in the simplified spherical shell with its boundary conditions (continuity of the tangential electric field and of the perpendicular displacement current density). The large variations of the boundary conditions imply that iterative numerical methods for the solutions have to be utilized. It turns out that an assumption of total reflection, i.e. infinite conductivity in the boundary surfaces of the shell, is a good starting point.

Tables of the variation of the earth's conductivity over different areas are given in Chapter 3. These show variations of approximately 4–5 orders of magnitude. In areas with low conductivity the ground moisture is of greater importance because the real part of the total dielectric constant then has a relatively larger influence. At the same time the skin depth gets larger, with the result that it is necessary to consider the conductivity computed to an average of about 100 m when calculating the propagation of OMEGA signals.

As mentioned in Section 4.3.3.1, different modes of propagation can exist in the spherical waveguide (as in any waveguide). There are different field configurations and different phase and attenuation constants for each mode. At the receiver antenna the mode voltages add vectorially, and large errors can arise if the high-order modes are of the same order of magnitude as the first mode.

Extensive computations of the attenuation and phase constants have been carried out using the methods indicated in Wait and Spies, 1964. Figures 4.6, 4.8 and 4.10 have been selected as representative examples of the results. Measurements show reasonably good agreement with the theoretical values.

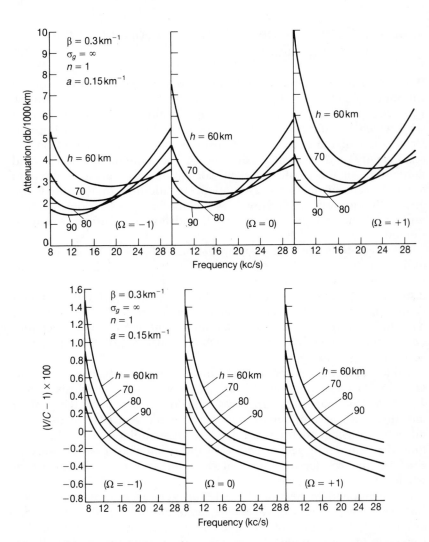

Figure 4.10 Attenuation and phase velocity as functions of frequency with the reflection altitude and the magnetic field as parameters. Negative values of Ω imply propagation to the east, positive ones, propagation to the west (Wait and Spies, 1964)

4.4.4 Receivers

An OMEGA receiver measures the phase difference between signals
from at least three stations for every position determination. To do
this, the receiver has to select the best stations and synchronize its
own signal segment generator with their signals. One method is to
utilize an accurate radio synchronizing signal and to utilize the fact
that the transmitters start their sequences at integer numbers of tens
of seconds. Another is to examine the strength of the received
signals, assuming that the strongest signal comes from the closest
station. However, this is not a safe method as anomalous wave
propagation may imply that more distant stations give a higher field
strength at the receiver. Besides, a certain focusing at the antipode of
a transmitter always occurs, especially for the eastward wave, such
that receivers near this point may receive stronger signals than at
points closer to the transmitter. For this reason it is recommended
not to use signals near their focal point. Automatic synchronization
means that the locally generated signal sequence in the receiver is
correlated to the received sequence until maximum output signal is
obtained from the correlator. The integration time is several minutes.
The difference in segment length (Figure 4.2) is important here.

The best and fastest synchronization method utilizes the identification frequencies introduced (12.1 kHz for the Norwegian station,
etc.), as these are specific to each transmitter. However, only the
newest receivers are equipped to do this.

The phase measurement itself, which is carried out after synchronizing, requires a memory of some kind as a certain frequency is used
by only one transmitter at a time (Figure 4.2). In ships, simple
reference oscillators can be used as the ship velocity is smaller than 1
per cent of a lanewidth in 10 s. In aircraft, on the other hand, more
sophisticated receivers are used, taking the larger velocity into
account.

Increase of the non-ambiguous area (Equation (4.7)), requires use
of more than one frequency. In the very simplest receivers which are
able to receive 10.2 kHz only, the lanewidth is only about 15 km
(8 n.m.). Receivers for more than one frequency utilize the frequency
difference between the signals. Then, the lanewidth becomes half a
wavelength at the difference frequency. The safest method is to solve
the ambiguity step by step, i.e. a receiver for 10.2, $11\frac{1}{3}$ and 13.6 kHz
makes its first lane identification at $(11\frac{1}{3}-10.2)$ kHz and its second
identification at $(13.6-10.2)$ kHz instead of going directly to 10.2 kHz
after 1.133 kHz. The corresponding half-wavelengths are 132 km,
44 km and 15 km respectively.

At these freqencies, atmospheric noise (lightning discharges, etc.) dominates over receiver noise. For this reason, nothing can be gained by using large antennae as the signal-to-noise ratio is constant in the received signal. A whip antenna with a length of between 2.4 and 4.5 m is common on board ships, whereas crossed-field antennae (two orthogonal loops) with outer dimensions of about $20 \times 25 \times 5$ cm are used in aircraft. The latter antennae are made to receive the horizontal, magnetic field component of the signal. Such an antenna is also advantageous with regard to reduction of static discharge and precipitation noise.

4.4.5 Transmitters

As indicated in Table 4.1 there are two main types of transmitting antennae: valley or fjord spans, and towers. The latter are either earthed at the bottom with wires of about 500 m lengths as radiating elements in an umbrella-like construction at the top, or isolated with a radiating tower. The tower antennae may be more than 400 m high. The valley spans are long wires, the transmitter antenna at Bratland, for example, consists of wires of 3600 m stretched between two mountains on each side of the Alder strait, and the lowest altitude over the fjord is about 400 m. The transmitter antennae in the OMEGA system have a bandwidth of only 10–30 Hz, in order to be reasonably efficient (15–20 per cent). Without this bandwidth restriction it would have been possible to transmit just one modulated carrier instead of a number of individual frequencies for lane ambiguity resolution. Each transmitted frequency has individual antenna tuning. The radiated power of the transmitting stations is 10 kW.

4.4.6 Accuracy

The dominating source of error in OMEGA is due to the propagation conditions (Section 4.4.3). The maps of the lines of position of the system are valid for average transmission conditions, but those not satisfied with accuracies of 10–30 km may improve accuracy by correction data published by the US Naval Oceanographic Office and others. These data are given as tables, maps or programmed corrections. They consider ionospheric properties, propagation direction and earth conductivity. The tables or maps divide the earth into areas within which corrections can be regarded as constant. The corrections are also given as functions of time of day and year and may in this way also consider, for example, SID and PCA to the extent that these can be predicted. Without such corrections, SID may give errors of 0.1–5.0 n.m., PCA 0.1–16.0 n.m. and westward passage of the magnetic equator 0.5–6.0 n.m. Uncorrected errors of this type may thus cause lane errors, even in an already synchronized receiver (lane slippage). This phenomenon may also occur when correction data are used (as all ionospheric fluctuations cannot be predicted), but in this case with smaller probability. Figure 4.11 shows the probability of these errors compared with errors during stable conditions.

Initially, published correction data were based totally on theoretical calculations. The most important sea areas have since been surveyed

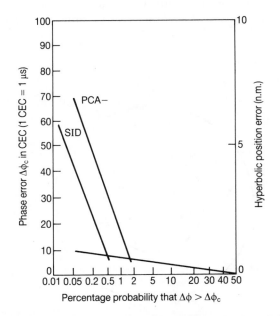

Figure 4.11 Percentage probability that the position error exceeds a given value under the influence of PCA and SID. 3.8 centicycles (600 m) is the estimated variance of a certain OMEGA signal in stable conditions (Swanson, 1977). The original version of this material was first published by the Advisory Group for Aerospace Research and Development, North Atlantic Treaty Organisation (AGARD/NATO) in *Conference Proceedings CP 209 – Propagation Limitations of Navigation and Positioning Systems* in February 1977.

systematically. It has thus been possible to reduce the errors for users in these areas by utilizing measured corrections.

Long-term measurements at many sites have shown that the OMEGA errors, after correction, are largely found within 1–2 n.m. rms in the day time and 1–3 n.m. rms at night (Figure 4.12). In certain areas of the earth, however, it may be difficult to drop below 5 n.m. rms, e.g. when using signals which have passed Greenland or the Antarctica.

The OMEGA Navigation System Center (Washington DC) is operated by the US Coast Guard. Among other tasks this centre continuously monitors signals crossing the polar areas. One of these specially observed paths is from Hawaii to Norway which is well calibrated and utilized for PCA detection. (Other such paths are Japan–North Dakota and Argentina–Australia.) A PCA is declared if the signal from Hawaii received in Norway advances its phase by at least 20 centicycles during more than 2 hours at the same time as the phenomenon is confirmed by other observations. In such cases, users

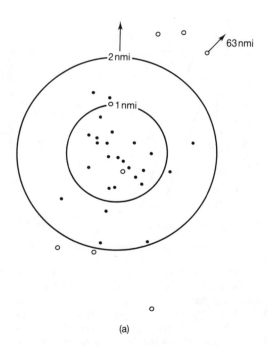

(a)

Figure 4.12 OMEGA position errors, after correction, in California and New York, (a) measured 1967–69 at 10.2 kHz

are recommended through broadcast messages to exercise caution in using OMEGA navigation solutions.

It was shown in Section 2.2.1.2 that the position accuracy in a hyperbolic system is inversely proportional to the sine of the angle between the intersecting lines of position.

Because of the even distribution of OMEGA transmitters over the globe it is always possible to find stations whose position lines intersect by angles of 60–90°. This makes it possible to keep the multiplication factor of the phase errors at 2 or better in most cases.

4.4.7 Differential OMEGA

As the irregular propagation conditions vary quite slowly with position, a large part of the error is approximately the same within a small area. (Tables from the Oceanographic Office are divided into areas of about 240 × 240 n.m.) This can be utilized by measuring the position at the known site, computing the error and then transmitting

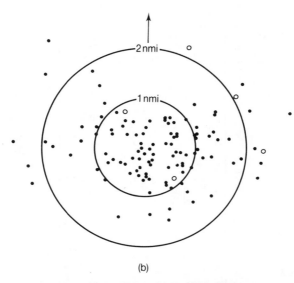

(b)

and (b) at five sites all over the USA in May 1969. Dot means normal conditions, ring means PCA (Swanson, 1977). The original version of this material was first published by the Advisory Group for Aerospace Research and Development, North Atlantic Treaty Organisation (AGARD/NATO) in *Conference Proceedings CP 209 – Propagation Limitations of Navigation and Positioning Systems* in February 1977.

Table 4.2 Differential OMEGA transmitters and coverage in July 1983

Name of station	Country	Code	Frequency (kHz)	Of beacon	Identifi- cation	Latitude	Longitude	Range (miles)	Correct trans.	Operational from
New and recommended stations 1 to 8										
Yeu Creach	France	9/0	312.6	CW-A2	YE	46°43'05" N	2°22'55" W	270	ABDEFH	6-75
(Ouessant)	France	10/6	308.0	SEQ-A2 1/6-H+5	CA	48°27'33" N	5°07'39" W	270	ABDEFH	4-78
Porquerolles	France	11/3	313.5	SEQ-A1 2/6-H+4	PQ	42°58'59" N	6°12'21" E	400	ABDEFH	12-76
Lagos	Portugal	12/0	364.0	CW-A2	LGS	37°09'33" N	8°36'52" W	500	ABCDFH	2-78
Porto Santo (Madeira)	Portugal	13/0	338.0	CW-A2	PST	33°03'44" N	16°21'25" W	500	ABCDFH	2-79
Horta (Açores)	Portugal	14/0	380.0	CW-A2	FIL	38°31'13" N	28°41'18" W	400	ABCDF	2-79

Table 4.2 *Cont.*

Name of station	Country	Code	Frequency (kHz)	Of beacon	Identification	Latitude	Longitude	Range (miles)	Correct trans.	Operational from
New and recommended stations 1 to 8										
Cap Bon	Tunisia	15/7	313.5	SEQ-A1 2/6-H+0	BN	37°04'13" N	11°02'34" N	400	ABCDEFH	3-80[1]
Gris-Nez	France	16/5	310.3	SEQ-A2 1/6-H+4	GN	50°52'10" N	01°35'01" E	150	ABCDEFH	5-80
Galantry (Saint-Pierre)	France	17/0	342.0	CW-A2	Y	46°45'56" N	56°09'17" W approx.	300	ABCDF	8-80
Port-Bouët (Abidjan)	Ivory Coast	18/0	294.2	CW-A2	PB	05°14'N approx.	03°56' W approx.	300	ABCDEF	1-81
Dakar	Sénégal	19/0				14°43'N approx.	17°28' W approx.	300	ABCDEF	[2]
Cayenne (Guyane)	France	20/1	327	A1	FXC	04°49'31" N approx.	52°21'54" W approx.	300	ABCDFG	[3]
Pointe-à-Pitre (Guadel.)	France	21/1		A1				300	ABCDFH	[3]
Cabo Finisterre	Spain	22/2	310.3	SEQ-A2 2/6-H+2	FI	42°52'51" N	9°16'17" W	270	ABDEFH	3-82
La Isleta (Canaria)	Spain	23/7	219.9	SEQ-A2 2/6-H+0	LT	28°10'18" N	15°25'08" W	300	ABCDFH	4-82
Cabo de Palos	Spain	24/2	294.2	SEQ-A2 2/6-H+2	PA	37°38'06" N	00°41'23" W	150	ABDEFH	4-82
Puerto Rico	U.S.A.		298	CW-A2	X	17°59'24" N	65°53'07"	150	ABCDFGH	8-82[4]

(1) Mode of operation not definitely determined

(2) (3) 1984 project

(4) Experimental station

Source: Nard, 1989.

this on a special communication channel to other users in the area. These users are then able to correct their own measurements. The method, called *differential OMEGA*, may reduce position errors considerably.

For many years, French companies and authorities have carried out systematic operations to map the possibilities of differential OMEGA. Continuous coverage now exists with reference transmitters from the southern part of the North Sea down to the Bay of Guinea in Africa, in the western half of the Mediterranean, in the Caribbean and off the Canadian east coast (Table 4.2). Corrections are transferred as modulations of the signals from maritime radio beacons in the areas.

In 1979 the predecessor of IMO (International Maritime Organization) passed a recommendation with regard to specifications of the differential OMEGA (IMCO, 1980). These specifications maintain, among other things, that

- The correction transmitters should have frequencies in the range 250–500 kHz.

- A subcarrier of frequency 20 Hz should be utilized for corrections of the 10.2 kHz signal and $23\frac{1}{3}$, $26\frac{2}{3}$ Hz and 30 Hz, respectively, for corrections of the signals at 11.05, $11\frac{1}{3}$ and 13.6 kHz (if those frequencies are transmitted, only 10.2 kHz is compulsory).

- The corrections are transmitted during one dash of the signal from the radio beacon.

- The subcarrier phase modulates the carrier to the modulation index <0.6, and the phase of the subcarrier directly corresponds (degree by degree) to the phase correction to be transferred.

- Updating is made at least every sixth minute and usually every third minute.

- It should be possible to correct the signals of at least four OMEGA stations.

- The corrections should have a resolution of 0.01 lane (about 150 m).

- The accuracy should be ±0.5 n.m. (95 per cent) within 100 n.m. from the reference transmitter (the radio beacon).

- The range of the beacon is usually determined by international agreement but should preferably be at least 200 n.m.

Measurements of differential OMEGA using the published geographic (absolute) corrections (Section 4.3.6) have given a relatively linear increasing error as a function of distance (Figure 4.13), starting at 0.3 n.m.

4.4.8 Prospects

Initially, expectations were high for the utilizability and accuracy of OMEGA. The problems of varying propagation conditions, however, appear to be more difficult to solve than was originally assumed, so the system has not achieved its expected popularity. It is now mostly used over large open sea areas where no other navigation system is available. The present worldwide number of users is about 12,000 (air) and 7500 (surface) (Federal Radio Navigation Plan, 1988). In Norway, some reconnaissance aircraft and helicopter traffic in the North Sea use OMEGA. Accuracy is, however, not always satisfactory; and in aircraft, particularly in helicopters, static discharges cause trouble and disturbance in bad weather (Forssell, 1982).

According to the US Federal Radio Navigation Plan (FRNP, 1988) the US authorities retain OMEGA in operation well into the next century.

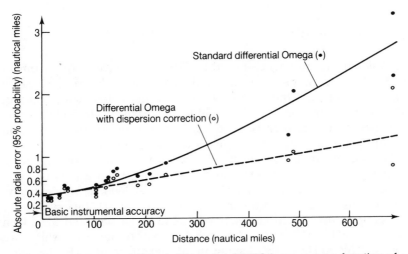

Figure 4.13 95% probability of differential OMEGA errors as a function of distance is 0.2 n.m. at 50 n.m. distance, 0.4 n.m. at 100 n.m., 1 n.m. at 500 n.m., etc. (Nard, 1989. Courtesy M. G. Nard, Sercel, and La Direction du Service Hydrographique et Océanographique de la Marine, France)

4.5 DECCA

4.5.1 Introduction

DECCA is a terrestrial hyperbolic navigation system whose stations transmit continuously in the frequency range 70–129 kHz. The transmitting stations are arranged in so-called chains consisting of a main station (master) with control functions and three (in some cases two) slaves whose signals are phase locked to those of the main station. The system is British and was introduced during World War II. It has been used mainly in Europe where most of the coastal waters are covered, but also in Japan, India, Pakistan, the Persian Gulf, South Africa and parts of Australia and Canada (although some of these regions are no longer covered). Consequently, it is a very widespread area radionavigational system, and in 1987 there were 140 stations in 42 chains in 17 countries (Beattie, 1988). In Norway there are six chains, Skagerak, Vestlandet, Trøndelag, Helgeland, Lofoten and Finnmark. DECCA is chiefly used by ships and, to some extent, by aircraft as well, especially helicopters. Trials on land have also shown fairly good results, both in the United Kingdom (Powell, 1982) and Norway.

4.5.2 Frequencies

The fundamental frequencies used by DECCA, called f, vary from 14 to $14\frac{1}{3}$ kHz and characterize the individual chain. The master station transmits at 6f, and the slaves, which have the colour codes purple, red and green, transmit at 5f, 8f and 9f, respectively. In addition, the frequency 8.2f (orange) is used for transfer of identification and control signals, and in some receivers for navigation also.

The fundamental frequencies f are distributed according to a nominal separation of 180 Hz (at 6f) and have been given the code designation 1B, 2B, etc., for 6f = 84.280, 84.460 kHz respectively. Some chains deviate from this pattern by 5 Hz (at 6f); this applies to, for example, 0B which has 84.105 (instead of 84.100) and 3B which has 84.645 kHz. There are also so-called half-frequencies which are separated by 90 Hz (at 6f) having the designations 0E, 1E, 2E, etc., for 6f = 84.195, 84.370, and 84.550 kHz respectively. In addition, the letters A and C are utilized for frequencies 5 Hz below and above the B values, and D and F for 5 Hz below and above the E values respectively. The frequencies 5A–F thus mean $6f = 84.995$, 85.000, 85.005, 85.085, 85.090 and 85.095 kHz.

4.5.3 Receivers

The position determination is based on measurement of phase differ-ences between the signals from the main station and the slaves. As these signals are transmitted at different frequencies, the receiver has to find a common comparison frequency. In analogue receivers this is achieved by multiplication of the received frequencies so that the phase difference between the master and the red signal is measured at $24f$, between the master and the green signal at $18f$ and between the master and the purple signal at $30f$ (Figure 4.14). These frequencies then determine the width of a lane, which is the range within which the phase difference varies less than 2π. At $18f$ this width is about 590 m, at $24f$ it is about 440 m and at $30f$ about 350 m (on the baseline).

The so-called *multipulse method* is used for lane identification. During the multipulse, each transmitting station in turn transmits all four frequencies ($5, 6, 8$ and $9f$) simultaneously for 0.45 s in a certain time interval, while the other transmitters of the chain are silent (Figure 4.17). This is repeated every 20 s. As the transmitted frequen-cies of the multipulse are multiples of the fundamental frequency f, the period of the composed signal is $1/f$. This allows a receiver to deduce the phase of the (non-transmitted) $1f$ signal and hence determine its position in a $1f$ hyperbolic pattern (rather than $18f$, $24f$ and $30f$). The received sinusoidal signals and the sum are shown in Figure 4.15.

The advantage of the multipulse method is that it is relatively insensitive to phase errors in the received signal components as these errors tend to cancel each other. This is shown by Figure 4.16(a)–(c) where different phase shifts of the received components change the shape of the curves within the period but where the position of the peak of the multipulse waveform is not changed. The least favourable combination of phase error signs allows a phase error of as much as 33° to occur in each component before a false peak, equal in amplitude to the wanted peak, intrudes and the system fails (Figure 4.16(c)).

The signal format for DECCA is shown in Figure 4.17. Every 20 s the stations transmit multipulse signals in the sequence MRGP, together with a signal component at $8.2f$.

A zone is the area within which the phase difference measurement

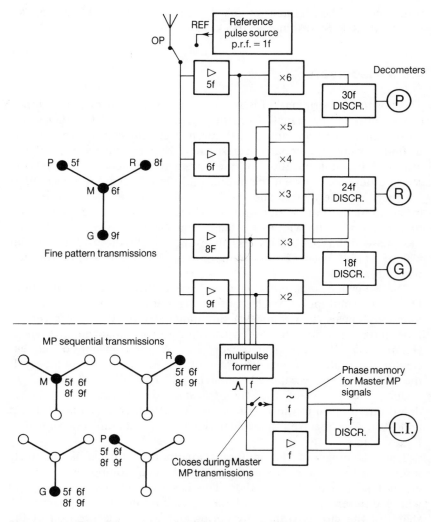

Figure 4.14 The principle of a DECCA receiver with lane identification by means of the multipulse method (DECCA, 1979. Courtesy Racal-DECCA Marine Navigation, Ltd)

(at 1f) gives an unambiguous result, and it has a width of $\lambda/2$ at 1f, i.e. 10.5–10.7 km on the baseline. One zone is 24 lanewidths at the red frequency, 18 at the green frequency, etc. The zones have the letter designations A–J, starting from the master. Receivers having zone identification measure phase differences at the difference fre-

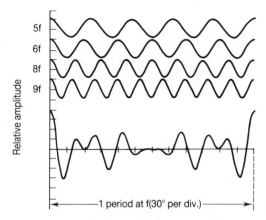

Figure 4.15 The four sinusoidal signals which form one multipulse and the resulting signal shape. Even if the sum signal has many zero crossings in one time interval of the length of $1/f$, there are relatively sharp pulses at the interval boundaries (DECCA, 1979. Courtesy Racal-DECCA Marine Navigation, Ltd)

quency 8.2f–8.0f, thus giving an unambiguous area of five zones (called a group). These areas are designated AF, BG, CH, DI and EJ in a DECCA receiver.

Analogue receivers used to be equipped with a watch-like display, a so-called Decometer, for each of the slave colours (Figure 4.18). In addition to hands giving the position in the individual lane with a resolution of 0.01 lanes, zone letters are also displayed. The two hands are coupled mechanically, and every zone corresponds to a whole 360° turn of the long hand (24 lanes for a red Decometer, 18 for a green, etc.). The short hand describes one turn for each lane. The Decometers are resolvers with perpendicular field coils and sine and cosine signals from the phase detectors as input signals.

Position fixes, with the help of Decometers, are obtained by plotting the readings in a lattice grid. Marine charts overprinted with the DECCA grid have been published by the hydrographic agencies of many nations. Modern receivers give the position data as longitude and latitude on digital display. Today's microprocessor controlled receivers (for DECCA as well as for other navigation systems), analogue as well as digital receivers, also have many other computing functions built in. The user can, for example, obtain a visual presentation of distances and directions to selected positions (way-points), his velocity, signal levels and quality (S/N) from different transmitters, etc.

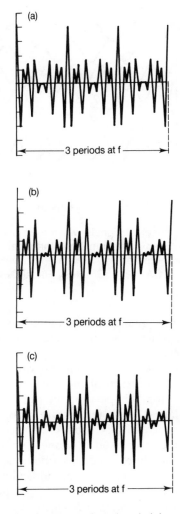

Figure 4.16 Three periods of a multipulse signal: (**a**) no phase shift in the received signal components; (**b**) +16° for 5f and 8f but −16° for 6f and 9f (this is the most favourable sign combination); (**c**) +33° for 5f and 8f but −33° for 6f and 9f. In (**c**) there are unwanted peaks of the same amplitude. (The time unit on the horizontal axis is 30° at 1f.) (DECCA, 1979. Courtesy Racal-DECCA Marine Navigation, Ltd)

To illustrate some receiver principles, a receiver introduced some years ago (Last and Linsdall, 1984) has been chosen. It has only one receiver channel which is common to all frequencies (i.e. time

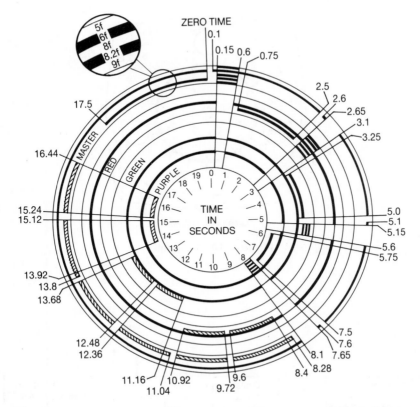

Figure 4.17 A signal sequence of 20 s for a DECCA chain. The stippled parts are signals at 8.2f, above all utilized for transfer of data and control signals, and chain surveillance (DECCA, 1979. Courtesy Racal-DECCA Marine Navigation, Ltd)

multiplexing). This is implemented by means of a switchable filter at the receiver input (Figure 4.19). A microprocessor controls the switching to the different frequencies by means of field effect transistors switching capacitors of the filter in or out. The passband of the filter can be centred around any of the five frequencies of each of the sixty-three DECCA channels. The local oscillator frequencies are controlled in synchronism with the centre frequency of the filter. The received signal is mixed down to a fixed frequency of 1.99 kHz (Figure 4.20). There are two LO-frequencies; the first one mixes the received signal down to the frequency f. At the other intermediate frequency (1.99 kHz) the signal passes a limiter, thus obtaining a square wave shape, and is then fed to the phase discriminator. This is

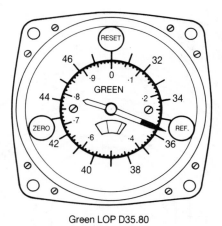

Green LOP D35.80

Figure 4.18 Decometer

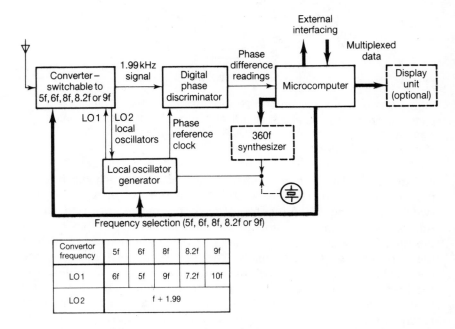

Frequency selection (5f, 6f, 8f, 8.2f or 9f)

Convertor frequency	5f	6f	8f	8.2f	9f
LO 1	6f	5f	9f	7.2f	10f
LO 2	f + 1.99				

Figure 4.19 Block diagram of a microprocessor controlled receiver (Last and Linsdall, 1984. Courtesy The Institution of Electronic and Radio Engineers)

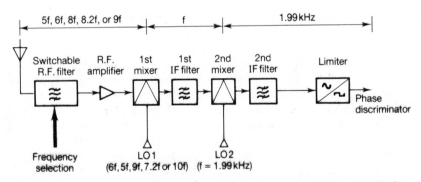

Figure 4.20 Frequency switching in the receiver above. Received DECCA signals at the frequencies 5f–9f are converted to 1.99 kHz (Last and Linsdall, 1984. Courtesy The Institution of Electronic and Radio Engineers)

in reality a counter with the frequency 1.99 MHz, and the phase detection implies reading the state of the counter into the micro-processor, triggered by an edge of the square wave (Figure 4.21).

The receiver operates in synchronism with a chain and follows its signal pattern of 20 s (Figure 4.17). When the receiver is switched on,

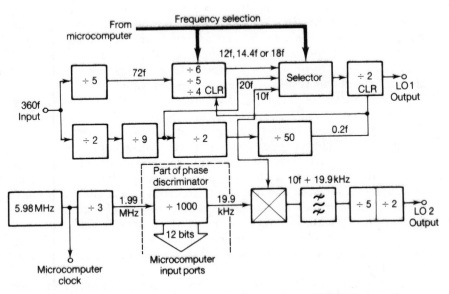

Figure 4.21 LO synthesizer and part of a phase discriminator (Last and Linsdall, 1984. Courtesy The Institution of Electronic and Radio Engineers)

it begins to search for the master signal, finds the pattern and then synchronizes to the whole signal format.

Between the multipulses the phases are measured sixteen times for each of the four main frequencies, and in the multipulses the receiver is switched rapidly between the frequencies, measuring the phases four times at each. By means of a computing algorithm with the same characteristics as a second order phase-locked loop, measurement results are averaged. Thus the effective receiver bandwidth is reduced from 30 Hz (the bandwidth of the input filter) to about 0.03 Hz.

The computation of the lines of position differs from that of a multiplying, analogue receiver described above. The phase measurements at the different frequencies during the multipulses are combined, first by means of the multipulse algorithm to find the correct lane at frequency f (one DECCA zone). This result is then utilized together with an optimization routine to resolve the ambiguities at each signal frequency. Finally, the phase differences at the signal frequencies are utilized in order to find the correct line of position. In order to extend the unambiguous area, the signal frequency 8.2f–8f is also used, giving a lane width of about 53 km.

The multipulse algorithm is based on the creation of lane fractions from the master–slave phase differences measured at 5f, 6f, 8f and 9f. The task of the algorithm is to find the zone fraction (at f) which best fits the four lane fractions measured. This is achieved by testing successive zone fraction values. For each, the corresponding lane fraction values at the four frequencies are computed, followed by a computation of the discrepancies between them and the measured values. The sum of these four discrepancies is at a minimum when the zone fraction being tested is closest to the true value.

One advantage of this type of receiver is that the rapid switching between the different frequencies makes receiver drift negligable. The use of a common receiver channel for all signals has a similar influence. In addition, pilot tones (i.e. receiver-generated signals at the transmitter frequencies), are transmitted at certain intervals through the receiver to control the phase delays of the receiver itself. The effective receiver bandwidth of 0.03 Hz gives a good S/N and, hence, good accuracy of the phase measurements (Equation (4.2)). Measurements with this receiver in summer night conditions have given results showing a repetition accuracy of 2.5–5.7 centilanes (68 per cent) of the lines of position at the centre of a chain (Last and Linsdall, 1984).

A recent receiver design also uses the single-channel time-multiplex principle. As indicated by the block diagram in Figure 4.22, the receiver can be regarded as having an analogue and a digital part. In

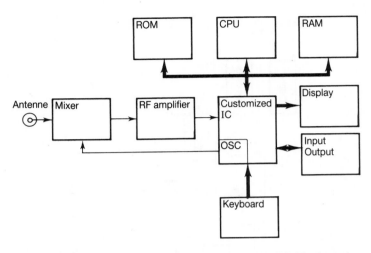

Figure 4.22 Receiver block diagram (courtesy AP Navigator)

the analogue part, the antenna coupler contains a quadrifrequency bandpass filter (Figure 4.23) which is also an impedance transformer from the whip antenna to the front-end amplifier. After amplification the different signals received are mixed in turn with a synthesized LO-signal 32 kHz offset with regard to the different frequencies. The IF-signal is hard limited and its phase is compared with a local reference (Figure 4.23).

All the four DECCA transmissions received from the chain in question are phase-measured for 0.5 s each, 0.5 s also being used for multipulse reception. Thus there is a complete updating of the phase information every 2.5 s. The time interval is locked to the master signal.

All phase measurements are referred to the fundamental frequency (1f) of the chain which is also used as a reference for position computation. This computation, which utilizes the method of least squares, also produces an estimated position error which is used for control of the automatic chain shift, so that the optimum chain is selected.

4.5.4 Transmitter stations

The transmitting antenna is usually a tower about 100 m high. The power of the transmitter itself is 1.2 kW at each frequency, but because of the short antenna ($0.02–0.04\lambda$) the radiated power is only

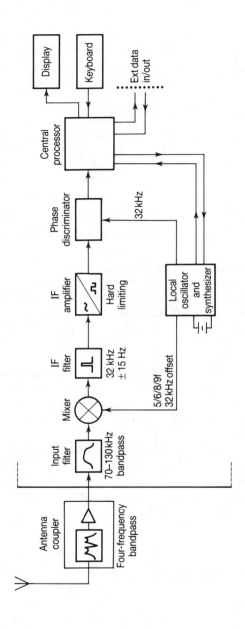

Figure 4.23 Single-channel time multiplex (courtesy AP Navigator)

100–200 W at each frequency. The chain is controlled from the master which has a stable oscillator at 6f. Oscillators in the slaves are normally phase-locked to this signal. In addition, the slave transmitters contain a controllable, fixed phase shift which is adjusted so that the master and slave signals are in phase on the extension of a baseline (taking the frequency difference into consideration).

4.5.5 Maps and corrections of the measurement result

Maps have been published showing lines of position for DECCA chains and the designations as described above, i.e. zone letters and lane numbers of chains and transmitter pairs. The maps of course contain designations of the chains so that the receiver's reference oscillator can be tuned to a correct frequency. With automatic receivers this is performed automatically after the user has entered an approximate position. The frequencies of the chain and certain corrections are stored in read-only memories in such receivers.

Because of irregularities of the propagation conditions (see following section), there may be systematic errors in the lines-of-position maps. When a chain is established, the lines of position are computed from the sites of the transmitter stations and calculated values of the propagation constants. These computations can be made reasonably accurately for propagation over sea (Chapter 3), but for signal propagation over land, measurements are usually needed to determine the values. The propagation velocity minimizing the variance of the measured values is chosen for the computation of the lines-of-position maps, but there may be local deviations, mainly due to differences in mean propagation speed as between the transmission paths to the observer from the master and slave stations of a pair. Because of this, corrections have been published to be used in addition to the lines of position, and Figure 4.24 shows one example of these.

4.5.6 Accuracy and coverage

The accuracy of the DECCA system is, to a large extent, dependent on the position of the user with regard to the transmitting stations, the time of year and the time of day. Even if the distances from the receiver to the transmitters are well within the stated range of the system, the accuracy varies because of propagation condition changes.

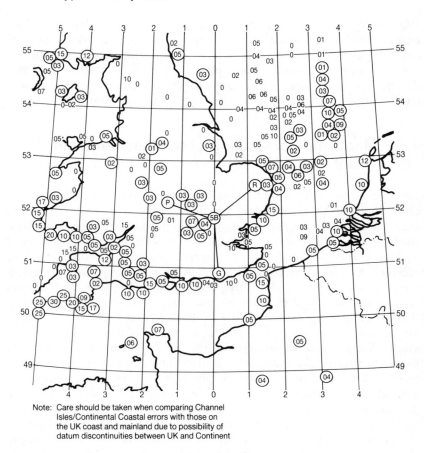

Note: Care should be taken when comparing Channel
Isles/Continental Coastal errors with those on
the UK coast and mainland due to possibility of
datum discontinuities between UK and Continent

Figure 4.24 An example of a map of systematic errors of a DECCA chain
(DECCA, 1979. Courtesy Racal-DECCA Marine Navigation, Ltd)

One part of the error can be said to be systematic because it is due to
an erroneous assumption with regard to the time average of the
propagation velocity of the signals in the respective direction. This
error usually varies with position, so over a large area there exists a
mean and a standard deviation. Thus in that case even this error may
be regarded as random.

The standard deviation of DECCA position errors is often counted
in hundredths of an average lanewidth (called a *centilane*), and this
unit is defined to be 5 m (DECCA, 1979). This may be a suitable unit
when navigating within a chain, the signals of which have a relatively
short propagation path, over sea water only. In the case of unfavour-

able terrain conditions, e.g. signal propagation over a mountainous ground with low conductivity (for example, ice formed by snow and fresh water), a unit ten times larger may be more suitable.

In addition to systematic errors due to irregularities along the transmission path there may also be errors in the lines-of-position maps due to the coordinate system itself or its reference. Other examples of systematic errors with similar magnitude are calibration errors in the chain.

Examples of random errors at a specific measurement are those caused by short time variations of the ionosphere, phase noise in the receiver and reading errors. The Racal-DECCA company has stated that the standard deviation of such errors is 5 m at each individual measurement of distance differences on summer days when the largest distance to the transmitter is 275 km and the transmission path as a whole goes over sea water (DECCA, 1979).

The standard deviation of the above type of errors, σ, is not totally independent of the distance to the transmitter. The distance dependence has been stated to be (DECCA, 1979)

$$\sigma = 5 + 0.003(1 + d) \ (\text{m}) \tag{4.9}$$

where d is the distance to the midpoint of the baseline normalized to 100 n.m. ($d = 1$ corresponds to 100 n.m.). In winter σ increases by a factor of three or more.

The Gaussian distribution gives the best description of how random errors appear in position determinations by means of DECCA in the day time and not too far from the transmitter.

The standard deviation of 5 m mentioned above is an average for all three colour coordinates of DECCA on the baseline. This implies that in Equation (2.20) $\sigma_1/2 = \sigma_2/2 = 5$ m, and when $k = 0$ the radius of the error circle is, accordingly,

$$R = \frac{5}{\sin \gamma} \sqrt{\frac{1}{\sin^2 \frac{1}{2}\alpha} + \frac{1}{\sin^2 \frac{1}{2}\beta}} \ (\text{m}) \tag{4.10}$$

When a DECCA receiver is used on or in the immediate neighbourhood of land, position errors may be caused by, for example, mountains, bridges, trees, high buildings and power lines (Reynolds, 1953; Sherman and Johnsen, 1976–7). The reason for this is that such objects scatter the signals such that the measured phase of the signal in the receiver does not reflect the shortest propagation path from the transmitter. In addition, the signal can be attenuated with a reduced signal-to-noise ratio as a consequence. Based on experiences and measurements it is recommended that the receiver should be at a distance from the protruding object of at least three times the height of that object (Powell, 1982). If the distance is smaller than that height, there may be large errors. Power lines in addition radiate noise and interference signals, particularly in damp or cold weather because of corona, and this may be a reason to increase the security distance.

Position determination based on red, green and purple fine pattern transmissions may give unsatisfactory results on land because of errors in determination of lanes or zones, caused by large and irregular variations of phase and amplitude. The multipulse, on the other hand, is often considerably more resistant to such errors (Figure 4.16) and may give usable results. Trials in the city of London have shown that the standard deviation of multipulse measurements in large cities is around 0.2 mean lanewidths (i.e. 100 m) (Reynolds, 1953).

DECCA is also used in aircraft to some extent, particularly in helicopters. Scattering in mountainous terrain often causes fluctuations of the lines of position with periods from a few tenths of seconds to several seconds, depending on the terrain and the flight velocity. (A helicopter flying 240 km/hour takes about 22 s to cover half a DECCA wavelength.) These effects can be mitigated to a large extent or even removed by filtering, in the receiver or in the display.

Aircraft use, including helicopters, may give static electric discharges on the fuselage and wings or rotors, increasing the noise level considerably. At the antenna input such interference effects may be several tens of dB above the level of the wanted signal. In such cases the receiver may be blocked for several minutes, especially in bad weather (Fanneløp, 1982).

Signals at DECCA frequencies are reflected by the ionosphere. At average latitudes and distances in summer, however, the reflection coefficient is small, about 2 per cent, and the effective reflection height is about 70 km. At night, however, the effective reflection height increases, thereby increasing the attenuation because of the increased propagation path length, but this effect is not enough to balance the attenuation of the signals in the lower parts of the ionosphere in the day time. The sky wave is therefore stronger at night than in the day time at average and large distances. The reflection coefficient is about 25 per cent at 500 km of distance at night, and the effective reflection altitude is about 95 km (DECCA, 1979).

The range of a chain is usually defined with regard to the distance where the wave reflected from the ionosphere reaches the same level as the ground wave, which is about 440 km at night and about twice that value in the day time (Figure 4.25).

Measurements of ground waves have shown that their magnitudes are approximately Rayleigh distributed. Even if the ground wave and the sky wave have about the same magnitude at about 800 km distance from the transmitters at night, there may be strong and rapid fluctuations of the received signal at shorter ranges, when the two waves have opposite phases. This can cause lane slippage when a sky wave on average has only half the power of the ground wave. This may happen about 400–500 km from the transmitters. The range where the sky wave and the ground wave are about equal is about 500–1300 km from the transmitters. After 1300 km the sky wave dominates. The DECCA lines-of-position maps have been worked out based on the ground wave alone. Sky wave interference gives fluctuating phase errors because of amplitude and phase variations of the sky wave. This gives rise to complicated error functions in the position measurement because the slaves are phase-locked to the master. Therefore, there are three different sky waves with influence at each line-of-position determination: master–slave, master–receiver and slave–receiver (Appendix 6). The statistical distribution of the errors can be predicted and, for example, be expressed as the part of the time the errors are below a certain value. The sky wave does not give rise to errors in lane counting until it dominates the master or the slave signal. This happens, as mentioned above no closer to the transmitters than 400–500 km at night (800–1000 km in the day time).

It was mentioned above that line-of-position deviations have turned

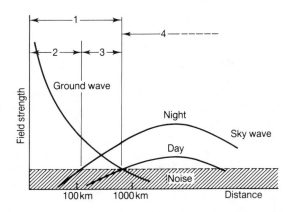

Figure 4.25 Field strength as a function of the distance for ground and sky waves: (1) daytime range; (2) night range (ground wave); (3) unstable range (fading); (4) sky wave alone

out to have a Gaussian distribution. Approximately, this applies also at night even if the measured density function has turned out to have a slightly different shape as about 75 per cent of all measurements are found within $\pm 1\sigma$ (compared to 68 per cent for the Gaussian distribution) (DECCA, 1979). This also implies that fewer measurement errors are found far from the average, compared to what can be expected according to the Gaussian distribution.

Extensive measurements of sky wave effects have to be carried out in order to make a good prediction. Such measurements have been carried out wherever DECCA chains are established, at different places within the chain coverage and at different times. Consequently, it has been possible to map variations as a function of times of year, meteorological phenomena, geographical position and earth conductivity. In order to avoid too much influence by short time effects, the measurement series have often been divided into time intervals of seven days.

The meteorological effects seem to be correlated to the sun spot activity which has a period of about 11 years. The uncertainty brought about by this activity is 10–20 per cent around the average, which the error predictions are based on, and is greatest in winter. The DECCA errors are at a minimum when the activity is at a maximum, and this is due to the attenuation of the sky wave under such circumstances.

The largest error sources within the coverage of the chain are irregular propagation and unfavourable aspect angles between lines of position. Maps have been published for the different chains showing accuracy contours (1σ) as a function of time of year and time of day (Figures 4.26–4.27).

4.5.7 Perspectives

As mentioned above, DECCA is widely used as a coastal navigation system, and the number of users is large (about 155,000 marine users in 1990 (Beattie, 1988)). In Norway, DECCA is the navigation system with the largest number of receivers in use. During the last few years many receivers, reasonably priced and easy to use, have appeared on the market. This has resulted in wide use of the system, even in small pleasure boats. Because of the inherent properties of the system with regard to accuracy and coverage, it is, however, natural to expect that such users will also turn to other, more accurate, navigation systems when such systems become available at acceptable prices.

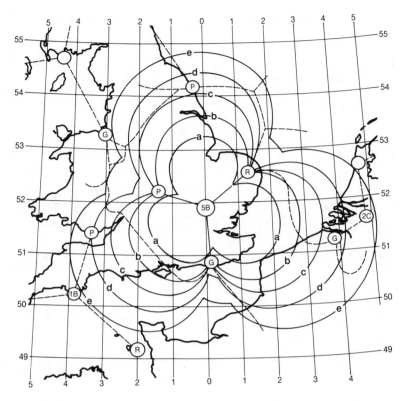

Figure 4.26 Decca-chain 5B (England). Coverage and accuracy (68% confidence) for other times than full daylight. Accuracy contours a, b, c, d and e are explained and quantified in the table and the onion diagram below. (Decca 1979. Courtesy Racal-Decca Marine Navigation, Ltd)

There is no expansion of DECCA chains at present, and it has been proposed in Norway that the Norwegian chains should be closed down if and when the LORAN-C system is expanded to full coverage of Norwegian waters (Section 4.6.10). In 1987 the continued running of the UK DECCA chains became the responsibility of the General Lighthouse Authorities. The contract is for seven years (i.e. until 1994), extendable until 1997 (D'Oliveira, 1988). (Decisions in April 1990 indicate that it will not be prolonged, and that DECCA will be replaced by LORAN-C after 1997.) Some other countries, e.g. Sweden and The Netherlands, want to keep DECCA as a national system for many years.

RANDOM FIXING ERRORS AT SEA LEVEL IN NAUTICAL MILES 68% PROBABILITY LEVEL

DECCA PERIOD See Time and Season Factor Diagram below	CONTOUR					
	a	b	c	d	e	
HALF LIGHT	<0.10	<0.10	<0.10	0.13	0.25	
DAWN/DUSK	<0.10	<0.10	0.13	0.25	0.50	
SUMMER NIGHT	<0.10	0.13	0.25	0.50	1.00	
WINTER NIGHT	0.10	0.18	0.37	0.75	1.50	

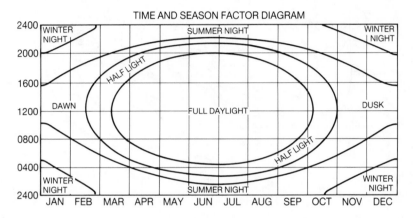

Figure 4.27 Rms errors at sea level (n.m.) and variations as function of time of the day and time of the year (Decca 1979. Courtesy Racal-Decca Marine Navigation, Ltd)

4.6 LORAN-C

4.6.1 General

The hyperbolic system LORAN-C (LOng RAnge Navigation) has been developed from an earlier navigation system (LORAN-A), also based on pulsed signals, which was invented in the United States during World War II. The first LORAN-C chain became operative on the east coast of the United States in 1958. From 1959 The Norwegian Sea chain has had stations at Ejde in the Faeroe Islands (master), Jan Mayen, Bø (southwest of Tromsø in northern Norway), Sylt (on the northernmost part of the German North Sea coast) and at Sandur in western Iceland. At present there are around fifteen LORAN-C chains covering the Mediterranean, the northwest Atlantic, the waters around Hawaii and Japan, southeast China, the whole west coast of North America (including Alaska), the regions around the great lakes in North America and down to the Gulf of Mexico. There are also two chains in Saudi Arabia, and so-called minichains in other parts of the world, e.g. the Suez canal. In addition, the system is being expanded to cover the rest of continental United States, mainly in order to serve air traffic. This expansion will be complete by the end of 1990 (Heyes, 1988).

The chains consist of a main station (master, M) plus two, three or four secondaries (X, Y, Z and possibly W, also called X-ray, Yankee, Zulu and Whiskey respectively). There are also four Soviet chains (Westling, 1984), one in the central European part of the country with five transmitters, one on the Pacific Coast with five, and two newly established chains with three transmitters each covering the western Arctic Ocean region of the Soviet Union. The Soviet system, called Chayka (seagull), has a signal format so similar to the US chains that the Soviet and US built stations can be utilized simultaneouly by some receivers (Westling, 1984). The baselines, i.e. the distances between transmitters of the same chain, are about 1000–1200 km.

4.6.2 Signal format

The system is based on time as well as phase difference measurements. The transmitters utilize the same carrier frequency, 100 kHz, but transmit pulses in time sharing with certain intervals (Figure 4.29). Coarse measurements are used to find the approximate times of arrival of the pulses and then the phase of the carrier is measured to obtain a better accuracy.

The transmitted pulses should have the standardized shape shown by Figure 4.28. The envelope of the pulses can be written approximately:

$$v(t) = v_0 \left(\frac{t}{t_p} \right)^2 \exp \left[2 \left(1 - \frac{t}{t_p} \right) \right] \qquad (4.11)$$

where $t_p = 65$–$70 \ \mu s$ (usually $65 \ \mu s$). The transmitted spectrum is required to have 99 per cent of the pulse energy within the frequency range 90–110 kHz.

The signals from each transmitting station contain groups of pulses,

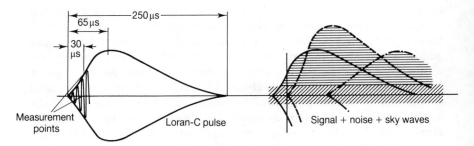

Figure 4.28 Loran-C pulse shape

with eight pulses in each group and 1 ms between the pulses in the group. In addition, the master transmits a ninth pulse, 2 ms after the eighth in its group (Figure 4.29).

The ninth pulse exists mainly for historical reasons, as it was previously used to identify the master signal by means of an oscilloscope; but its task is also to transfer information on abnormal conditions in the signals of the secondaries. This is made by blinking the ninth pulse at intervals of 12 s, transferring the letter groups RE,

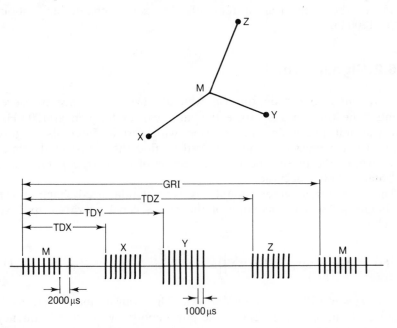

Figure 4.29 Signal format of a LORAN-C chain. Each line means a pulse of the type shown in Figure 4.28

REE, REEE and REEEE in Morse code to indicate that the stations X, Y, Z and W, respectively, transmit unusable signals. The secondaries themselves also announce such conditions (if possible) by blinking the first two pulses of each group, on for 0.25 s and off for 3.75 s.

Each chain is characterized by the period of these pulse groups, GRI (Group Repetition Interval, Figure 4.29). There are forty possible GRIs, and the GRIs of neighbouring chains have to be chosen with great care so that signals from different chains do not overlap (Section 4.6.3). The pulse group intervals (Figure 4.29) are chosen so that signals from the same chain do not overlap either. The Norwegian Sea chain has GRI = 79,700 μs, i.e. there are 79,700 μs between each period of master transmission. (The Icelandic chain, as a comparison, has 99,800 μs.) The GRI expressed in tens of μs (i.e. 7970 for the Norwegian Sea is often used as designation of a chain.

As all transmitters utilize the same carrier frequency, those which belong to the same chain must transmit in a fixed sequence. Previously, the master signal received at the secondaries activated these, and after a fixed delay, CD (Coding Delay), they transmitted their own signals. Today, the CD may vary slightly as part of the control procedure of the chain (see below). Behind the secondary, as seen from the master, the Time Difference, TD, between the master and secondary signals is equal to CD. On this baseline extension, the smallest TD values of the coverage area are found. The largest TD values are found on the baseline extension behind the master as seen from the secondary where

$$TD = 2MS + CD \qquad (4.12)$$

(MS is the travel time of the signal between the master and the secondary.) The time between the two transmissions from the transmitters, ED (Emission Delay) is then

$$ED = MS + CD \qquad (4.13)$$

The 7970 chain has nominal values as shown by Table 4.3. It is obvious that there is no risk of overlapping signals within the chain as $(TD)_{max}$ of a preceding pulse group is much less than $(TD)_{min}$ of the subsequent one.

However, the chains are no longer run with fixed CD values. Instead, the secondaries are controlled by local cesium clocks (and not by the master signal), which, of course, may drift a little in relation to the master clock. MS varies in any case and the chain

Table 4.3 Nominal values of time delays (μs) for chain 7970

	$(TD)_{min}CD$	MS	$(TD)_{max}$
X(Bϕ)	11 000.00	4048.16	19096.32
W(Sylt)	26 000.00	4065.96	34131.32
Y(Sandur)	46 000.00	2944.47	51888.94
Z(Jan Mayen)	60 000.00	3216.20	66432.40

control is organized in such a way that TD is measured by a fixed monitor station in a central area of the chain. If a measured value exceeds a certain tolerance interval, a message is sent to the respective secondary, which then adjusts its CD stepwise with an integer number of tens of nanoseconds. In this way TD at the monitor station is kept fixed to a theoretically computed value with little variation allowed, independent of the variations in propagation conditions for the signals (Doherty and Feldman, 1975). It may thus be said that LORAN-C is run differentially, with monitor stations as references. But as the propagation variations are not equal in the whole of the coverage area, there will always be smaller or larger errors in other parts of the area.

One frequency is used at each transmitter, so the phase of the carrier and the pulse intervals are coherent.

In order to reduce the influence of the interference and to enable the receiver to identify the signals automatically, these are phase coded (Frank, 1960). The phase code implies that the carrier of the transmitted pulses starts either by a positive ($+$) or negative ($-$) edge. (The pulse in Figure 4.28 is thus a plus pulse.) The master code is $++--+-+-+$ and $+--+++++-$ which is transmitted every two times (thus with GRI in between). Correspondingly for the secondaries (all alike), $+++++--+$ and $+-+-++--$. With this code, signals reflected from the ionosphere with delays corresponding to one or more pulse intervals have little influence on the detection (Section 4.6.4). The codes are also of great importance to the search procedure in the receivers (Sections 4.6.5 and 4.6.6). The codes used are complementary Golay codes. The codes of the master and the secondaries are orthogonal.

4.6.3 Interference and use of pulse groups

Interference by other signals is sometimes a great problem, especially in Europe (de Bruin and van Willigen, 1988). The sources are found inside the LORAN band (e.g. LORAN transmitters in other chains

(Zeltser and El-Arini, 1985)), or outside (e.g. DECCA and broadcasting transmitters). One often distinguishes between synchronous, near-synchronous and non-synchronous interference, defined as

$$\left| f_{int} - \frac{m}{2GRI} \right| \begin{cases} = 0 \text{ (synchronous)} \\ \leq f_b \text{ (near-synchronous)} \\ > f_b \text{ (non-synchronous)} \end{cases} \qquad (4.14)$$

where f_{int} is the interference frequency and f_b is the receiver tracking bandwidth (usually 0.01–0.5 Hz), m is an integer.

The probability of near-synchronous or synchronous interference is reported to be in the order of 1 per cent (de Bruin and van Willigen, 1988). Such interference causes difficulties in locking on to the correct zero-crossing and is countered mainly by notch filters (Section 4.6.5).

The interference problems which may be created by the chain itself are of two kinds: skywave interference and interference from other transmitters of the same chain. The former problem is treated in Section 4.6.4. The latter problem has to be considered in connection with the design of the signal format of the chain and has been described in the preceding section. The transmitters have to operate in time sharing so that no pulses can overlap, irrespective of the position of the receiver, so that the identification of the signals is secured.

The necessary interval between the pulses from two transmitters is shown by Figure 4.30. The master transmits a pulse which arrives at the secondary after a time β. After that arrival plus the coding delay, Δ, the secondary transmits a pulse. Back at the master, the time

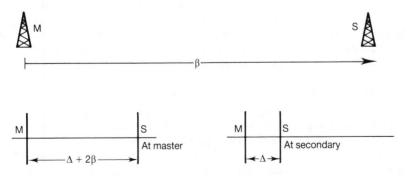

Figure 4.30 Time relations in a hyperbolic system with two transmitters

difference between the master and secondary pulses is, consequently, $2\beta + \Delta$. In order not to introduce overlap the interval between the master pulses thus has to be at least $2\beta + \Delta$.

The other problem, ionospheric interference, may also imply overlapping pulses. Multiple reflections, i.e. signals reflected several times by the ionosphere with ground reflections in between, may create pulse trains which at 100 kHz may have a duration of several tens of ms, even if the pulse amplitude is reduced at each reflection. If a second pulse is transmitted too soon after the first, multiple reflections from the latter may create errors in the envelope as well as in the phase of the carrier (Figure 4.31). In order to avoid this and to make time sharing possible in a practical way in LORAN-C, a buffer zone of at least 2 ms between subsequent pulses from different stations has been introduced (Figure 4.32). In the diagram, Δ is the buffer zone which is to prevent overlapping between the master and secondary pulses, and I is the time interval between the master pulse and the master identification pulse.

If, instead of transmitting one pulse, the master transmits a whole group of pulses, the effective pulse repetition frequency is increased. The necessary interval between these pulses is then less than Δ because they come from the same transmitter, and only the first pulse of each group has to wait for the reflections of the last pulse of the previous group to vanish (Figure 4.33). If the length of the group is G, the total pulse repetition interval is

$$GRI = 2G + 2\Delta + 2\beta + I \qquad (4.15)$$

Table 4.4 gives examples of values of the constants Δ, β and G as well as comparisons between the conditions in the case of signal pulses and pulse groups. The table shows that the number of pulses per second which can be transmitted from one transmitter increases when pulse groups instead of single pulses are utilized. This is very important because it reduces the requirement for peak power in the single pulses, as the necessary signal-to-noise ratio in the receiver can be achieved by coherent integration of several pulses instead.

In a LORAN-C receiver the integration time is limited by the requirement that the movement of the receiver has to be much smaller than one wavelength during the integration time. At 100 kHz the wavelength is 3 km, and the maximum movement during the integration should therefore be about 150 m. Ships with velocities of up to 30 knots can therefore use integration times of about 10 s, which is also a common value. As integrator in this case, a lowpass filter with 0.1 Hz bandwidth may be used.

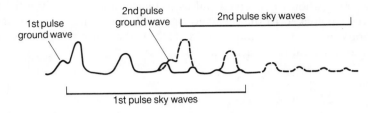

Figure 4.31 Multiple reflection skywave interference

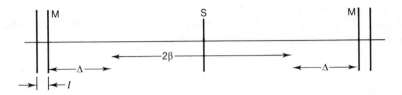

Figure 4.32 Interval requirements in the case of two master pulses and one secondary pulse

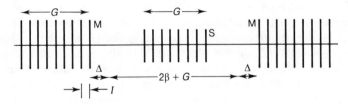

Figure 4.33 Intervals in grouping of pulses

Table 4.4 Improvement because of pulse grouping

	Single pulses	Groups-of-eight
2Δ	22	22
2β	9.096 32	9.096 32
I	2	2
$2G$	–	14
GRI	33.096 32	47.096 32
Pulses/s	$\dfrac{1}{GRI} = 30.2$	$\dfrac{8}{GRI} = 169.9$
Improvement factor	–	5.6 (= 7.5 dB)

Note: Numerical values in ms from MX-7970 transmitters.

The integration of received pulses is phase coherent only after the receiver has locked to the received signal. If, then, the received signal amplitude is A, the signal power is $A^2/2$. If the noise power is N, the signal-to-noise ratio of each pulse is

$$\left(\frac{S}{N}\right)_1 = \frac{A^2}{2N} \tag{4.16}$$

After phase-coherent integration of n pulses, the amplitude is nA and the power $(nA)^2/2$. The noise is, however, integrated incoherently as the noise in one pulse is statistically independent of the noise in another. The noise power after integration is, therefore, nN. Thus the signal-to-noise ratio after integration is

$$\left(\frac{S}{N}\right)_n = \frac{nA^2}{2N} = n\left(\frac{S}{N}\right)_1 \tag{4.17}$$

For the Norwegian Sea chain, where eight pulses come from each transmitter during 79.7 ms, the improvement factor at 10 s of integration is $10 \times 8 \times 1000/79.7 = 1000$ times, i.e. 30 dB. A signal-to-noise ratio in one pulse of -20 dB, which is sometimes given as a minimum requirement, then results in an effective signal-to-noise ratio of 10 dB, giving a measurement accuracy (rms) in each phase of $\sqrt{0.1}$ [rad] $\approx 18°$ (Equation (4.2)). If both the master and secondary signals are assumed to have the same S/N, and the noise is uncorrelated, the corresponding phase difference measurement error is 25° (Equation (4.3)). This corresponds to 0.07 lanewidths, i.e. 100 m.

4.6.4 Coded pulse groups

Phase coding as a means of reducing the influence from multiple reflections in the sky wave can be explained by Figure 4.34. The receiver knows the pulse interval and the phase code, and it compares the detected phase of the received pulses to a reference phase by sampling the received pulses at subsequent times. This sampling conserves the phase of the received signal. When the reference and the received phase have the same sign, a positive signal is produced, otherwise a negative one. As the sampling takes place only at the position of the reference pulses (i.e. with intervals of 1 ms), interference between the reference pulses has no influence. The code implies that two phase patterns are utilized so that every second group of eight pulses has one phase pattern, and the interleaving pulse groups

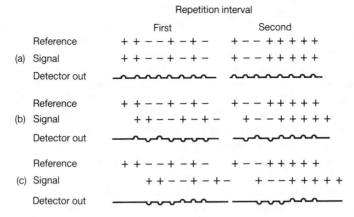

Figure 4.34 Cancellation of multihop skywave interference by phase coding of pulse groups (master signal)

have the second pattern (Figure 4.34). When the reference signal and the received pulse groups coincide (within one pulse width), the detector gives only positive outputs (Figure 4.34(a)), so that integrating the output signal of the detector gives a positive signal level. When the received signal is delayed one pulse interval (Figure 4.34(b)), there are equally many positive and negative detection pulses for two subsequent pulse groups, and the result of the integration is zero. The same thing happens if the received signal is delayed by two pulse intervals (Figure 4.34(c)). In this way the influence from multiple ionospheric reflections can be eliminated, something which appeared to be necessary during tests of LORAN-C in the 1950s. Then, non-coded signals showed variations in the effective propagation velocity of the signal which could be ascribed to multiple reflections with resulting delays of several pulse intervals.

4.6.5 Receivers

In LORAN-C, as in other navigation systems, there has been a development of receivers which has led to increasing utilization of microprocessors and considerably more digital signal processing. The description given below starts with conventional receiver principles which have been used for many years. After that, principles are described which are utilized in modern receivers with a larger computational capacity.

There are a large number of manufacturers of LORAN-C receivers, general receiver types as well as receivers for special purposes (MacKenzie, 1984).

The tuning of the receiver to the chain to be used is based on the GRI. The master signal is identified first. With the automatic receiver, pulse groups are generated with somewhat higher repetition frequency than that transmitted by the master, but with the same phase code, so that the local pulse groups pass slowly over the received ones (in time). Multiplication of the internally generated pulse train by the amplitude detected, received pulses, followed by integration (integration time up to about 10 s), gives a correlating receiver whose output signal reaches its maximum at the moment when locally generated and received pulses coincide. Then the internal GRI is adjusted to that of the chain so that the two pulse trains are interlocked. The principle is shown in Figure 4.35. Because of the phase coding the result after correlation is minimally influenced by sky waves with one or more pulse interval delays (Figure 4.34).

The polarity of sky wave products after the multiplier changes according to the phase code, such that the average value after integration of several pulse groups is well below the detection threshold.

When the receiver has locked to the master signals in this way, similar search procedures for the secondary signals are carried out as the sequence is known (Figure 4.29).

The receiver has a reference oscillator which is now phase-locked to the master signal, from which all other frequencies needed in connection with the reception are derived (GRI, etc.). The phase-lock must have a time constant so that the reference does not drift too much in the intervals where no master signal is received. This

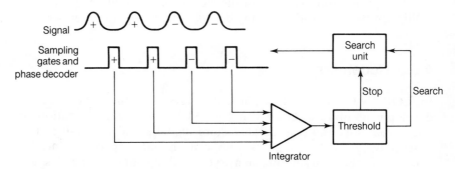

Figure 4.35 Correlating detection and lock-in of LORAN-C signals (only four pulses are shown to make it simple) (Gløersen, 1974)

also makes for a lengthy lock-in time. The total time from the start of the search until full synchronism is reached is usually several minutes.

During the search the receiver bandwidth is usually reduced to about 5 kHz from the normal 20–40 kHz in order to increase the signal-to-noise ratio. (On the other hand, this bandwidth reduction severely distorts the pulse shape, so it must not be used for the tracking receiver.) LORAN-C receivers often operate satisfactorily with S/N > −20 dB in the pulses. The reason for this is the long integration time (Section 4.6.3). During the search procedure, however, the integration is incoherent, and the improvement is therefore less, which is the reason for the bandwidth reduction.

As mentioned above, time as well as phase measurements are utilized in order to find a position. In the receiver the difference in time of arrival between the rising edges of the pulses from the master and secondaries is measured. This has to be done with an accuracy of ±5 μs as the period of the carrier is 10 μs. After that the difference is measured more accurately by measuring the phase difference between the carriers at the zero-crossings after three periods, i.e. 30 μs from the start of the pulses. The reason for this choice of time is that the time delay of the signals reflected from the ionosphere referenced to the ground wave is usually larger than 32–35 μs (Shapiro, 1968; Thrane and Røed Larsen, 1975), and the reason is consequently to avoid interference (Figure 4.37). In order to obtain a good margin to sky wave interference, it is common practice to track the zero-crossing after 25 μs.

The rising edge of the pulse itself is normally detected after 2.5 periods, that is, about 25 μs from the start of the pulse, where the pulse voltage is about half its maximum value. In order to find this point, in principle the anologue receiver utilizes two methods. In the first method it finds the difference between the envelopes of the received pulse and the delayed and amplified version of this, i.e.

$$v_1(t) = v(t) - A_1 v(t - \Delta t_1) \qquad (4.18)$$

(Equation (4.11) and Figure 4.36), where A_1 and Δt are chosen such that $v_1(t)$ is zero at the desired point. Using the other method, the received envelope is differentiated twice (Figure 4.38) so that the point of inflection

$$v_2(t_2) = A_2 \left[\frac{d^2 v(t)}{dt^2} \right]_{t=t_2} = 0 \qquad (4.19)$$

is the point of detection. (With the curve shape of Equation (4.11)

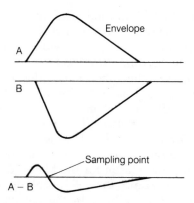

Figure 4.36 Detection of the rising edge of the pulse by the delay-and-compare method

this occurs about 21 μs after the start of the pulse so than an extra delay is needed even here.)

If the correct (i.e. third period) zero-crossing of the ground wave is utilized at the phase measurement, no signal should precede this by 30 μs; therefore the receiver is equipped with gates which check if this is really the case. If signals are found at that check point, the phase detection is moved to the zero-crossing before that in use, as the presence of signal power at the mentioned gate indicates that an erroneous (i.e. later) zero-crossing is being used.

At least in Europe, the most usual cause of wrong zero-crossing detection is interference from other transmitters (Section 4.6.3), in or near the frequency band used. To counter such interference, sharp notch filters are required in many cases. These filters, which can be automatically or manually tuned, cut off a part of the wanted signal, in addition to the unwanted one. This also implies pulse distortion. The width of the stop band of such filters is usually 1–2 kHz.

Because of the frequency dependent propagation conditions the different parts of the spectrum of the LORAN-C pulses are delayed differently on their way from the transmitter to the receiver. By this the phase is distorted, and the pulse shape more or less deviates from the nominal one (Equation (4.11)). This also displaces the point of detection on the rising edge, so that, in extreme cases, some other zero-crossing may be closer than the correct one (envelope-to-cycle discrepancy, ECD) (Jones, 1978).

In modern receivers most or all of the signal processing is done digitally. Rapid computational and control functions enable the

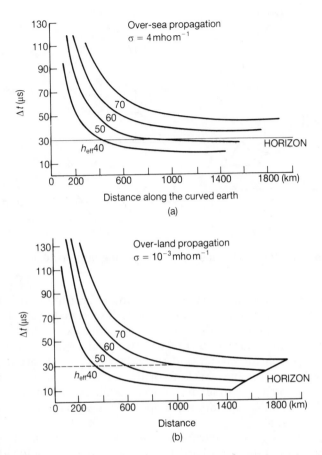

Figure 4.37 Computed time delay of the sky wave related to the ground wave for different reflection altitudes (one reflection) (Thrane and Røed-Larsen, 1975)

receiver, for example, to process signals in parallel from several transmitters (which means shorter search time among other things), automatic tracking of the different secondaries at the same time and automatic selection of the best transmitters (if more than one is available), based on signal strength and geometry. Of course, even such a receiver has the same principal functions as its predecessors, because the system is the same. All receivers can now compute and present the position as longitude and latitude, whereas older receivers only put out the time differences (TD) which had to be translated

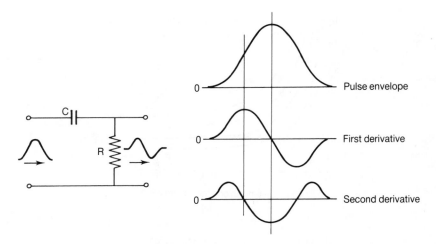

Figure 4.38 Detection of the rising edge of the pulse by differentiation

manually into position by means of maps overprinted with lines of position. Some new receivers are able to use signals simultaneously from different chains (cross-chain fixing) (Lamiraux, 1988).

The basic functions of a LORAN-C receiver are, as indicated above, the search for master and secondary signals, determination of the wanted zero-crossing, tracking of the envelope and the zero-crossing, measurement of time differences, addition of possible corrections and position computation (Poppe, 1982). A basic block diagram is shown in Figure 4.39.

Usually the search and tracking functions are separated, and the former often operates with hard limiting and reduced bandwidth (van der Waal and van Willigen, 1978). Also, the tracking function can operate with hard limiting (van Willigen, 1985). The disadvantages are mainly a higher sensitivity to cw interference and a slight reduction in S/N (about 2 dB for low values). Thus a hard-limiting

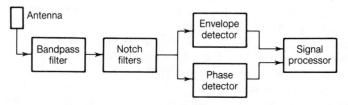

Figure 4.39 Basic block diagram of LORAN-C receiver

receiver is dependent on good notch filters before the limiter. The advantage of such a receiver is above all its simple construction and its robustness with regard to atmospheric noise (which often contains high-amplitude spikes).

Figure 4.40 shows an example of a block diagram of a microprocessor controlled receiver (de Lorme and Tuppen, 1975) operating in this way. (Even though this is not a very recent construction, it does show the basic functions very well.) During a search the limited signal is sampled in quadrature with 125 μs between the pair of samples (Figure 4.41). Through quadrature sampling, a sufficient signal amplitude at the sampling point is ascertained. At the same time the interval between the pair of samples is short enough to detect the presence of a pulse. The results from 8 + 8 samples with 1 ms interval are added after taking the sign of the wanted phase code into account (Section 4.6.6), so that the desired transmitter signal can be detected.

In order to find the desired zero-crossing, the signal is sampled at a number of points before the position which leads to the detection of the pulses. The interval between these samples is 40 μs (Figure 4.42). The detector jumps to an earlier point if a signal is present in the samples. The length of the jump is determined by the number of samples with signal present.

When the approximate position of the zero-crossing after the third signal period has been found, its position can be measured accurately and tracked. This is done by taking three samples with 2.5 μs (i.e. 90°) intervals around the zero-crossing of the signal from the linear part of the receiver (Figure 4.40). Adding the results from these three samples should give a total of zero if the mid-sample lies exactly on the zero-crossing and the two others are weighted with regard to the nominal shape of the pulse edge. (Two samples, one before and one after, would also add to zero, but an increased number of samples gives better performance in the presence of noise.) An algorithm gives computation and filtering of the sum of three so that the sampling points are directed towards, and kept, at the values which give the desired zero result. The procedure has the same characteristics as a phase-locked loop. The time of the mid-sample (Φ_{strobe} in Figure 4.43) is then the time of arrival of the LORAN-C signal, and this time should be compared with the corresponding time for other transmitters.

By means of sample-and-hold circuits (Figure 4.40) the envelope around the wanted zero-crossing is tracked. At the same time, the ratio between the amplitudes of the signal periods immediately above and below the zero-crossing are compared with the shape of the mathematically correct pulse.

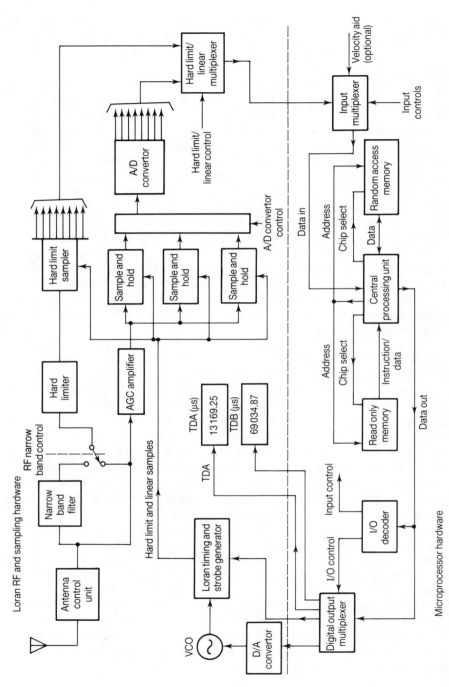

Figure 4.40 Block diagram of a LORAN-C receiver (de Lorme and Tuppen, 1975)

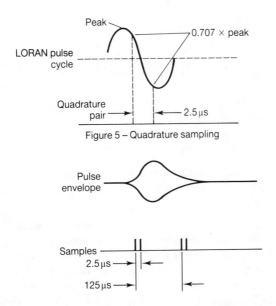

Figure 5 – Quadrature sampling

Figure 4.41 Sampling of a LORAN-C signal (de Lorme and Tuppen, 1975)

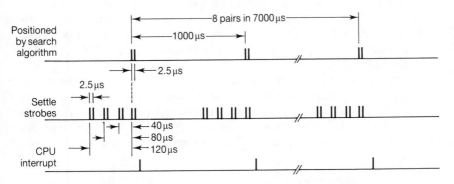

Figure 4.42 Sampling to find the wanted zero crossing (de Lorme and Tuppen, 1975)

4.6.6 Grouping of pulses for automatic search

To prevent erroneous locking by the receiver, search routines have been developed which exploit the phase coding in order to minimize the probability of locking to the wrong pulse, even in cases of strong interference (Frank, 1960). Then the pulse groups shown by Figure

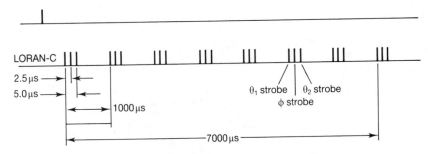

Figure 4.43 Sampling at phase tracking

4.44 are utilized as the receiver processes two groups together, but also divides each group into two (the ninth master pulse is not included). At the sampling, the received signal is multiplied by the first four pulses of the two subsequent reference pulse groups ($M1$) and the four last ($M2$), respectively. The result is a sampling of the received signal in intervals determined by the reference. The products are added, and finally $M1$ is multiplied by $M2$ and this product is compared to a threshold.

If the received signal is synchronized to the reference, each pulse in $M1$ and $M2$ gives a positive contribution at the sampling, i.e. totally $+8$ for each, but when the signals are not synchronous, the contributions are zero or negative such as indicated in Figure 4.34(b) and (c), and the sum is then <8.

Great care has been taken during the code selection for LORAN-C. Among other things, the changes from one pulse group to the next are made such that the 1st, 3rd, 5th, and 7th pulses always have the same phases, whereas the 2nd, 4th, 6th, and 8th pulses have a changing phase. This is shown for the master by Figure

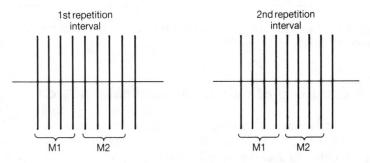

Figure 4.44 Subdivision of master pulse groups in receiver automatic search

4.34, and for the slaves by the two codes $+++++--+$ and $+-+-++--$ which follow the same principle. This implies that, in the correlation process which the synchronization really is, it is necessary to examine the correlation only for an even number of displacements of the pulse groups. This saves computing time as there are fewer states to examine when only every second pulse has a changing phase (see below).

In order to illustrate the correlation procedure, it is assumed that the conditions of Figure 4.34(c) are valid. The sampling then includes the reference pulses 3–8 (at the times where the reference pulses 1 and 2 arrive, there is no received signal). $M1$ is the sum of the number of even minus the sum of the number of odd signs of the first half of the two groups of reference and received signals, i.e. in this case only pulses 3 and 4 in the reference. These have the phases $--$ and $-+$, respectively, in the two reference groups. The phases of the signal pulses received at these times are $++$ and $+-$ respectively. A multiplication of the phases of the reference pulses by those of the corresponding signal pulses consequently gives the result $--$ and $--$. Thus $M1 = -4$. $M2$ includes the reference pulses 4–8 which in those two groups have the signs $+-+-$ and $++++$ respectively. The corresponding phases of the received pulses are as shown by Figure 4.34(c) $--+-$ and $-+++$, respectively. M2 is consequently $+4$. The product $M1 \times M2$, which is used as a measure of the synchronization, is -16, and as the receiver contains a threshold with a positive value, it notices that the signals are not in synchronism and continues to move the reference in relation to the received signal.

A detailed scrutiny of the phase multiplied signals for all even displacements between the reference and the received signals gives the results shown in Table 4.5. The threshold signal gets a large, positive value only when the synchronization is correct. If pure integration had been utilized, i.e. $M1 + M2$ instead of $M1 \times M2$, the correct reference position would not have been so well distinguished from the others (the rightmost column of the table). The non-linear method of synchronization described above is thus advantageous compared to linear signal processing which is used when tracking an already synchronous signal (coherent integration). An advantage of the chosen codes of the master and the secondaries is also that the cross-correlation is zero, irrespective of the combination of groups, and thus there is little danger that the receiver locks the master reference to the secondary signal or conversely.

In a practical noisy signal, of course, the conditions are somewhat different; cross-correlations, for example, are not exactly zero, and the threshold functions are less steep. Measurements have shown that

Table 4.5 Analysis of search procedure

Reference and received signal in the same pulse group	Ref. pulses no.	Samples* M1	M2	Threshold (M1 × M2)	Track val. (M1 + M2)
	1–2	0	0	0	0
	1–4	0	0	0	0
	1–6	−4	+4	−16	0
Master	1–8	+8	+8	+64 (Correct synchr.)	16
	3–8	−4	+4	−16	0
	5–8	0	0	0	0
	7–8	0	0	0	0
	1–2	0	0	0	0
	1–4	0	0	0	0
	1–6	+4	−4	−16	0
Secondary	1–8	0	0	0	0
	3–8	−4	+4	−16	0
	5–8	0	0	0	0
	7–8	0	0	0	0

Table 4.5 *Cont.*

	Ref. pulses no.	Samples* M1	M2	Threshold ($M1 \times M2$)	Track val. ($M1 + M2$)
Received signal displaced by one group interval	Master 1-2	+4	0	0	+4
	1-4	0	0	0	0
	1-6	-4	0	0	-4
	1-8	0	0	0	0
	3-8	0	-4	0	-4
	5-8	0	0	0	0
	7-8	0	+4	0	+4
	Secondary 1-2	-4	0	0	-4
	1-4	+8	0	0	+8
	1-6	-4	0	0	-4
	1-8	0	0	0	0
	3-8	0	+4	0	+4
	5-8	0	+8	0	+8
	7-8	0	+4	0	+4

Note: *summed over two pulse groups, each unit represents one pulse.

30–40 dB suppression of false synchronizing signals can be expected (Frank, 1960). This is of the same order as the suppression of multiple reflection interference by means of phase coding.

The search velocity (the time to reach lock), varies as a function of signal-to-noise ratio, and in the worst case it will be a few minutes. Before lock-in has occurred, the integration is incoherent, and the improvement factor is then less than the number of integrated pulses.

4.6.7 Transmitters

The most usual type of transmitter antenna, the top-loaded monopole (utilized by the Norwegian stations), is shown in Figure 4.45(a); the construction shown in Figure 4.45(b) is also utilized. Both antenna systems have large counterpoises (i.e. synthetic groundplanes); around them the radii are about 300 m and 500 m respectively. The antenna construction using the four towers is more expensive but has a considerably higher efficiency, which is important for the high powers used. The radiated power of a monopole antenna is normally about 200–400 kW (50–100 kW at the detection point on the pulse edge), whereas the same transmitter in the case of multitower antennae would be able to radiate about 1 MW of peak power. (The station at Sandur in Iceland radiates 1.5 MW because it is also part of the Icelandic chain.) Even in the latter case the efficiency is only about 10 per cent because of the large wavelength compared to antenna dimensions.

The power stage of older transmitters consists of four water-cooled triodes, two by two in parallel, operating in class B. They deliver an antenna current of 700 A maximum at an anode voltage of 21 kV (transmitter type AN/FPN-44A) (Sherman and Johnsen, 1976–7).

In modern transmitters the tubes have been replaced by semi-conductors. The transmitters utilize thyristors, capacitors and saturable reactors in so-called half-cycle generators (Figure 4.46). Every half-cycle generator generates half a period, positive or negative (not both), at 100 kHz with output power and starting time accurately controlled by means of feed current and time respectively. (The timing is controlled by a cesium clock.) The circuit of Figure 4.46b operates in the following way: first Q_2 is switched on so that C_1 is charged resonantly by a half-cycle lasting a relatively long time (175 μs) and with a low current. After that, Q_1 is fired by the voltage over C_1 so that the charging of C_1 is interrupted. Then C_1 sends its charge through the triggered Q_3 and through the resonant circuit $C_2 - T_3 - C_1$ in a half-cycle with a larger current (about 3 kA peak

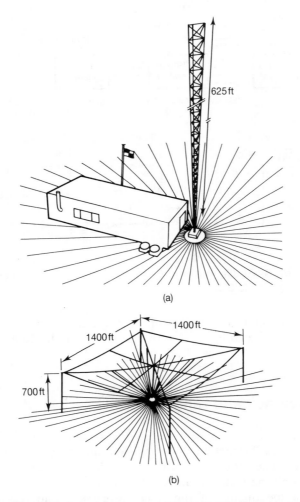

(a)

(b)

Figure 4.45 Transmitting antennas for LORAN-C (Sherman and Johnsen, 1976)

when T_3 saturates), which lasts for a shorter period (about 15 μs) than the preceding one. The half-cycle to the tank circuit of the transmitting antenna is made by discharging C_2 through T_5 and T_4 in a current of about 9000 A maximum in a half-wave which lasts 5 μs. Thus, succeeding stages of the half-cycle generator provide a shorter but larger half-cycle current pulse.

Phase coding is produced by firing a positive generator first for plus code phase, and a negative one first for minus code phase. The pulse

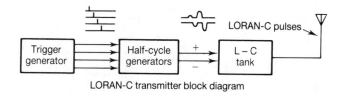

LORAN-C transmitter block diagram

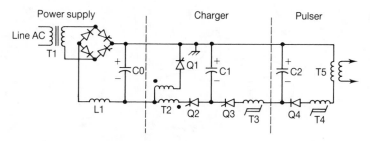

Figure 4.46 (a) Block diagram of a transmitter. (b) Half-cycle generator (Frank, 1983. © 1983 IEEE)

envelope shape and the relationship between envelope and cycles are controlled by regulating the currents of the individual half-cycle generators.

4.6.8 Accuracy and range

The range is defined in approximately the same way as for DECCA, i.e. according to the limits of position determination by means of the ground wave. The range, therefore, varies depending on the direction to the transmitting stations and the times of year and of day. It also depends on the quality of the signal processing and receiver. The ranges are usually considered to be 2000–3000 km over sea in the day time and about 30 per cent less at night. If the signals propagate over land, the range decreases by at least 10–15 per cent (Figure 4.47).

The largest source of error in LORAN-C is variation in the signal propagation velocity. The velocity of the ground wave depends on the earth's surface conductivity and (to a lesser extent) on the parameters of the atmosphere (Chapter 3). The range and the signal stability are therefore better for signal propagation over sea than over land. Every chain contains one or more monitor stations continuously measuring time differences (TD) of the signals of the different transmitters. (The Norwegian Sea chain, 7970, has a control station at Keflavik, Iceland,

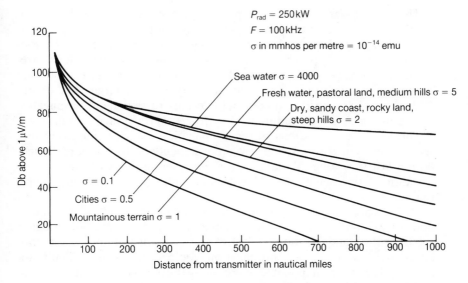

$P_{rad} = 250\,kW$

$F = 100\,kHz$

σ in mmhos per metre = 10^{-14} emu

Sea water $\sigma = 4000$

Fresh water, pastoral land, medium hills $\sigma = 5$

Dry, sandy coast, rocky land, steep hills $\sigma = 2$

$\sigma = 0.1$

Cities $\sigma = 0.5$

Mountainous terrain $\sigma = 1$

Db above 1 µV/m

Distance from transmitter in nautical miles

Figure 4.47 LORAN-C groundwave field strength

and, in addition, monitors on the Shetland Islands.) The control station has continuous links to all the transmitters, usually by dedicated full duplex telephone lines, and messages are given to the respective secondary when the TD has changed more than 20–50 ns with reference to the previous time of correction transmission. A secondary then changes its time of transmission stepwise with a multiple of 20 ns in each step so that the TD of the monitor station is kept approximately at its nominal value (Doherty and Feldman, 1975).

Even with such corrections, propagation variations reduce the accuracy as the correction from the monitor is exact only for its neighbouring area. Correction values of the TD within the coverge area have been measured. The most sophisticated receivers have such values stored in a read-only memory so that a measured value is corrected automatically as a function of position. Short-time fluctuations cannot be corrected in this way, of course, and a residual error is almost always present, even in the receivers, which can also be calibrated. Such a calibration implies that the position is measured at a known place where the user enters his position data. The receiver then computes the necessary correction factor of the propagation velocity in order to make the measured position correct, stores this factor and uses it (possibly scaled with regard to the distance to the transmitters) in further position determinations.

The largest variations of the propagation velocity are experienced over land and follow a periodic pattern with a period of one year. In areas that experience substantial changes in the properties of the earth's surface between summer and winter the variations are greatest. However, poor earth also has a stabilizing effect on the propagation velocity as a low conductivity gives a large skin depth (Equation (4.8)), so that a considerable part of the signal propagates at depths which are not influenced by frozen earth, precipitation, etc.

The long-time variations of the propagation velocity caused by the seasons are superimposed by weather-dependent fluctuations which do not last very long. Variations in the refraction index of the air as a function of the height over the ground have some importance here (Samaddar, 1980).

Much work has been put into theoretical and practical investigations of LORAN-C propagation (Samaddar, 1980; Wenzel and Slagle, 1983; Sæther and Vestmo, 1985). Although much information and knowledge is available, propagation factors are still difficult to predict. However, a possible error budget follows (all values in μs and 1σ):

Propagation anomalies, (over land)	0.2
Synchronization errors in secondaries	0.05
Tolerances in short-time variation data of propagation	2.0
User measurement errors	0.1
Geodetic errors and errors in the average propagation velocity for sea and land, respectively	0.1/0.5

Propagation short-time variation data may be much better than 2 μs. But the error usually lies between about 0.3 μs for optimum conditions and about 2.1 μs in cases of bad calibration and insufficient data. Corresponding position errors are 45 and 310 m respectively. In addition, we have the influence of angles of intersection of the lines of position. Figure 4.48 gives examples of resulting accuracies.

4.6.9 Differential LORAN-C

As the variations of the propagation velocity of the signals are caused by climatic changes, it is tempting to assume that they are correlated over larger or smaller geographical areas. It would then be possible to utilize this by measuring time differences at a place with a known position and transmit the deviations between the measurement result and the ·nominal value over a communication channel to other users in the area. These would then need to change their own measurement

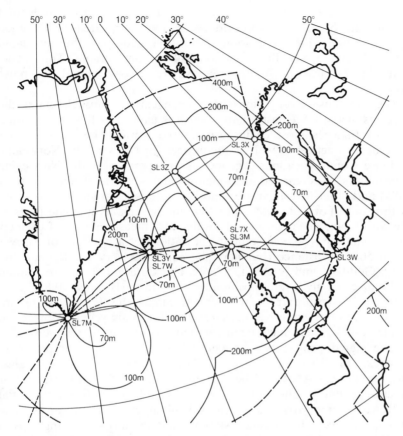

Figure 4.48 Calculated repetition accuracy of the chains in the Norwegian Sea (1σ). The limits of position determination by means of the ground wave are stippled (measured values usually better)

results by the received corrections in order to reduce the position errors correspondingly. The method, so-called differential LORAN-C, has been investigated very thoroughly and tested, particularly in the United States, and it has been found that it may result in considerable improvement of accuracy (Elias, 1985; Bruckner and Westling, 1985). However, it is important to be aware that only those errors which are correlated can be reduced in such a way. Examples of uncorrelated errors, which, consequently, are not reduced, are those caused by noise in the receivers. Bad signal-to-noise ratio is usually due to very distant transmitters. (Errors of this type are also geometry-dependent.)

Differential LORAN-C will be able to give an error reduction which can often be expressed as a ratio. In areas with good signal-to-noise ratios and good transmitter geometry the correlated errors dominate, and the reduction will be considerable. This reduction will be less in other areas where errors caused by noise dominate.

The previous investigations have been carried through under the auspices of the US Coast Guard, which is natural as this organization runs or finances most chains around the world. The investigations have been practical as well as theoretical. In Norway, the Norwegian Defense Communications Agency, as an operator of the Norwegian transmitters, has carried out measurements (Sæther and Vestmo, 1985) and other investigations, and has initiated further investigations in this area (Forssell, 1985). Practical measurements in some parts of the world have given very good results, e.g. measurements of the new French chain have given standard deviations of a few tens of ns over about 500 km distance (but here the earth has very good conductivity), (Goddard and Vicksell, 1986). The differential system around the Suez canal gives total position errors <15 m (Bigelow, 1983).

In the eastern part of the United States the US Coast Guard (USCG) has run extensive measurements for many years (Wenzel and Slagle, 1983). These have been carrried out to gain better knowledge of the signal propagation in general and to be able to compare practical and theoretical results. Experimental use of short-range differential LORAN-C has been conducted by USCG in several areas (Vichweg, 1988). The purpose is to be able to utilize differential LORAN-C in narrow waters, above all in harbour entrances. The idea is to establish reference stations and monitors covering only a small area, so that possible corrections eliminate the errors caused by variations in signal propagation parameters almost completely. The measurement results presented also show that very good accuracies can be expected from short-range systems, of the order of 8–20 m (Vichweg, 1988).

Another very important result given by the measurements in the United States is that the correlation distances of the long- and short-time variations mentioned above are very large or, put another way, the maximum correlation distance decreases very slowly as a function of stricter accuracy requirements. This implies that the reference station can cover a large geographical area if the requirements for accuracy are moderate. Expressed in figures it may be said that the reference station of differential mode, with a good location, can serve a whole chain with its corrections if the requirement for accuracy improvement referenced to the conventional mode is not stricter than 3–4 times (Forssell, 1985).

In order to transmit correction data from reference stations to users, the LORAN-C signals themselves have long been regarded as a possible means, and some practical tests have also been carried out. The advantages of transmitting the corrections by means of the navigation signals themselves are obvious, if the coverage areas are the same, which usually seems to be the case.

If the usual transmitted signals are used as information carriers, different types of modulation might be employed, which meet the important requirement that performance must not be reduced in common navigational receivers. The most attractive solution seems to be position modulation of one or more of the pulses in each group; this method is also the one that has been studied most thoroughly (Hoogenraad, 1988). The principle itself implies that the respective pulse is moved a certain number of μs related to its nominal position when a bit is to be transferred. In order not to disturb the function of the common navigational receiver, the pulse position must be changed very little and simultaneously be balanced, i.e. for any delayed pulse there must be a corresponding advanced pulse in the same group. Thus each bit requires an even number of pulses to be modulated (two, four, six or eight). In view of the low bit rates normally required, it is sufficient to utilize two pulses in each group. The last two pulses are most suitable as this will give the least changes in the behaviour of certain receiver filters.

4.6.10 Prospects

LORAN-C is a mature system and, as shown in the preceding section, it has the potential for very good accuracy. Its use in Norway has been limited by the coverage area, which at present includes only the waters off the coast of central Norway. An expansion of the system with new transmitters in the Bergen area and in the north-east near the Soviet border is in the process of determination but this depends on the north-west Europe expansion plans.

The US Coast Guard will withdraw its financial support of the operation of the chains outside the United States after 1994 (as a consequence of the operational status of GPS), and therefore a LORAN-C policy group composed of the countries of north-west Europe has worked out a proposal concerning the future of LORAN-C in this region. The solution will most probably be that each country with transmitters on its territory takes over the financing of their operation, possibly supported by the other countries. An

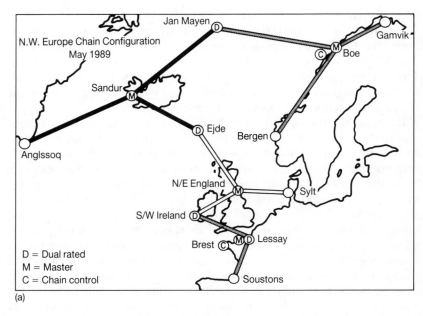

(a)

Figure 4.49 (a) Proposed configuration of LORAN-C in N-W Europe (Phase 1)

expansion has been proposed, consisting of new transmitters in Norway, the United Kingdom and Ireland. This also means reconfiguration of the chains, new chains (Figure 4.49) and refurbishment of the old transmitters (Stenseth, 1988). (Decisions taken in April 1990 indicate that this expansion will be realized as soon as possible.) A realization of these expansion plans means that the whole of north-west Europe is covered from the Bay of Biscay to the Arctic Ocean. Connections to the Soviet chains, both in the Baltic and in the Arctic are also being considered. That would be a natural continuation of the newly commenced joint operation of US- and Soviet-built transmitters in the northern Pacific (Westling, 1984).

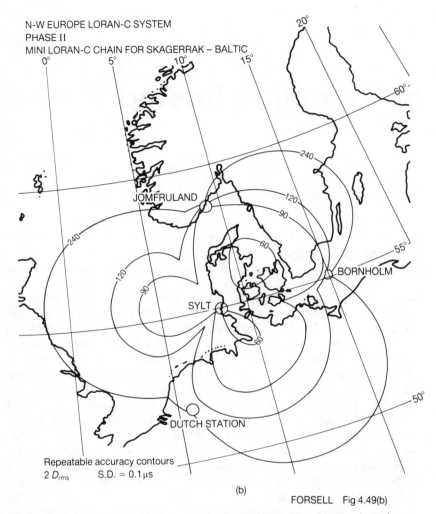

N-W EUROPE LORAN-C SYSTEM
PHASE II
MINI LORAN-C CHAIN FOR SKAGERRAK – BALTIC

Repeatable accuracy contours
2 D_{rms} S.D. = 0.1 μs

(b)

FORSELL   Fig 4.49(b)

Figure 4.49 (b) Proposed expansion of LORAN-C in the Skagerak and southern Baltic (Phase 2)

5 Direction finding

5.1 Introduction

Radio direction finding implies that the direction to a radio transmitter with a known position is determined by means of directional receiver equipment. By determining the directions to two different transmitters one's own position can be found by the crossing of the two lines of position (Figure 5.1). In an ideal system, the lines of position are great circles, as only the radio waves propagating along the shortest path from the transmitter to the receiver can be received.

If there is only one available transmitter, the position can be found by using the course and distance covered when the direction to the transmitter is determined from two different places en route. Knowing the course and the velocity of the ship, the position can be calculated (Figure 5.2). In this case the position accuracy is influenced by both the direction measurement errors at different points in time and by velocity and course errors. In other words, one is trading the baseline between two transmitters (in the two-transmitter case) with the vector between the measurement points (in the single-transmitter case) and the uncertainty in that vector.

Direction finding is the oldest and probably the most used method to find a position by means of radio waves. Special transmitters or radio beacons are utilized, as well as transmitters for broadcasting and other types of radio communication. Thus a number of transmitters can be used. The transmitter to be used is selected by the user in accordance with the capacity of his receiver and his approximate position.

Special transmitters sited along the routes are utilized for air traffic,

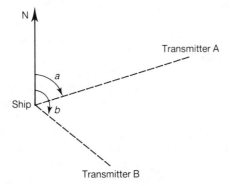

Figure 5.1 Position determination by direction finding

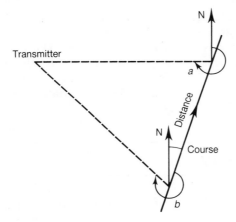

Figure 5.2 Direction finding using a moving line of position

so-called non-directional beacons (NDB), operating at MF. Direction finding principles are also used to determine the directions to aircraft as seen from an airport, as the transmitted signals from the VHF communication radio of the aircraft are used for the direction finding.

5.2 Frequencies and transmitters

The principle of direction finding is general and allows any frequency to be used. The frequency is selected according to the desired coverage of the transmitter, accuracy and international frequency allocations. The frequencies most utilized are in the range 0.2–1.7 MHz. The maritime and aircraft radio beacons usually transmit at 255–415 kHz.

The advantage of low frequencies like these is the coverage which is not limited by the horizon. During day time, ground waves dominate, but at night, far away from the transmitter, signals reflected by the ionosphere dominate. The disadvantage of low frequencies is antenna dimension, at the transmitter as well as at the receiver. The non-directional transmitter antennae are usually high masts, and in order to have good radiation properties these should be a quarter or half a wavelength. But the wavelength at 200 kHz is 1.5 km such that the antenna height is practically limited to fractions of the wanted dimension. To obtain better radiation properties, the antenna is made resonant at the frequency to be used but this introduces considerable losses, and the efficiency of the antenna is therefore often as low as 5–10 per cent.

Radiated power is of the order of 100 W. The radiated power and location of the transmitter station determine the range, which is usually 50–100 n.m. For obvious reasons, maritime radio beacons are located on the coast, whereas NDBs for aviation use are often sited at high-altitude inland locations.

5.3 Directivity

Low frequencies imply large antennae at the receiver also, if high accuracy is to be achieved. The directivity is caused by the fact that signals received by different parts of the antenna have different phases. The path length difference between the two antenna edges is $d \sin \varphi$ for a plane wave with incident angle φ (Figure 5.3), and the phase difference is consequently

$$\Psi = \frac{2\pi}{\lambda} d \sin \varphi \tag{5.1}$$

where λ is the wavelength and d is the antenna dimension. Close to the perpendicular a small change of the incidence angle, $\Delta\varphi$, gives the phase change

$$\Delta\Psi = \frac{2\pi}{\lambda} d \, \Delta\varphi \tag{5.2}$$

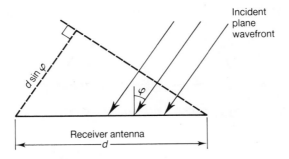

Figure 5.3 Reception of a plane wave

As shown in Section 4.2 the phase-measurement accuracy depends on the signal-to-noise-ratio such that

$$\Delta\Psi = \left(\frac{S}{N}\right)^{-1/2} \tag{5.3}$$

The phase difference between the signals from two-element antennae with identical receiver channels giving uncorrelated noise is then

$$\Delta\Psi_{1,2} = \left(\frac{S}{2N}\right)^{-1/2} \tag{5.4}$$

and the corresponding direction error is

$$\Delta\varphi = \frac{\lambda\sqrt{2}}{2\pi d(\sqrt{S/N})} \tag{5.5}$$

It should be noted, however, that at MF, atmospheric noise dominates completely over thermal noise, so the noise in the two channels is correlated.

5.4 Receiver antennae

As mentioned above the most common transmitter antenna is a vertical mast, so that the radiated signal is polarized vertically and the antenna has a toroidal radiation diagram (Figure 5.4).

As the receiver antenna always has to be small compared to the wavelength, loop antennae are often used (Figure 5.5). The voltage in

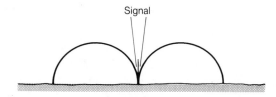

Figure 5.4 Vertical diagram of the antenna of a direction-finding transmitter

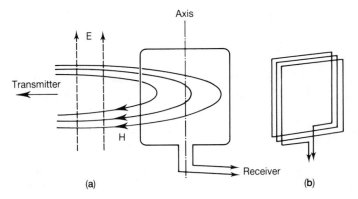

Figure 5.5 Loop antenna: (**a**) principle; (**b**) loop with several turns

a loop antenna is induced by the magnetic field. The antenna gain is proportional to the number of field lines inside the loop, i.e. to the area of the loop, the frequency and the number of turns.

The radiation diagram of a loop antenna is determined by the phase difference between the induced voltages in the side branches of the antenna (Figure 5.5). When the wave propagation direction is perpendicular to the plane of the antenna, the path length difference is zero, as is the output signal from the antenna. Thus the largest phase difference is given by a wave propagating in the antenna plane (Figure 5.6), and the antenna diagram is proportional to $\cos \Phi$ where Φ is the direction angle (Equation (5.1)) ($\Phi = 90° - \varphi$.) As shown in Figure 5.6, the diagram is circular, the reason being that the periphery angle with a circle diameter as a base is always $90°$, which makes the length of U in Figure 5.6 proportional to the cosine of the angle to the antenna plane.

Notice that the phase of the output signal from the antenna changes $180°$ from one circle to the other of the figure-of-eight. If a loop is turned, the signal voltage varies as shown in Figure 5.6, with sharp minimas when the perpendicular of the antenna plane points

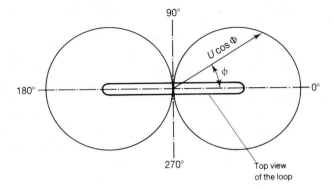

Figure 5.6 Radiation diagram of a loop antenna

towards the transmitter. In order to eliminate the ambiguity a non-directive antenna, e.g. a whip, is used, the signal of which is added, in phase, to the output signal from the loop. If the two signals have the same amplitude, the resulting radiation diagram is

$$V = V_0(1 + \cos \Phi) \tag{5.6}$$

i.e. a cardioid diagram (Figure 5.7). Here $|dV/d\Phi|$ has its maximum when $|\Phi| = 90°$.

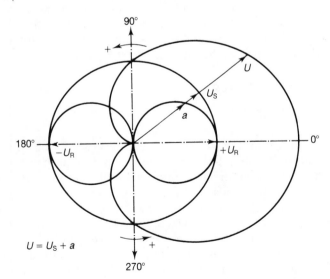

Figure 5.7 Principal radiation diagram of the receiving antenna

When the composed antenna is turned with regard to the direction of incidence of the received signal, the received voltage varies as shown in Figure 5.7. The sharp minimum shown by the diagram for 180° incidence direction is not always present. It requires, among other things, that the output voltages of the two antennae paths (non-directive and figure-of-eight) are exactly equal in the direction in which maximum occurs. In practical cases the cardioid is always more or less distorted, and it may be difficult to find the minimum point. The direction finding can then be made with a loop first to find the minimum direction to the transmitter (either 90° or 180° in Figure 5.7). The ambiguity is resolved by switching in the non-directional antenna in addition to the loop in the minimum direction found. If the sum voltage now increases with increasing turning angle of the whole antenna, 90° is correct (Figure 5.7); if it decreases, then it is 270°.

Due to mounting and space requirements of the revolving antennae, and to the requirement for mechanical stability, these antennae are often unpractical. Even at the beginning of this century it was found that two fixed, perpendicular loops could be utilized (Figure 5.8), a so-called 'goniometer' (*gonio* being Italian for 'angle'). The advantage of this arrangement is that the antennae can be mounted as high as possible, e.g. on board a ship, so that the reflection level is as low as possible. A balanced cable going from each of the two fixed loop antennae down to two equally fixed crossed-coils in the goniometer transfers only the voltages of the loop antennae down to the coils, the magnetic field of which then makes the search coil see the same field as if it were where the antennae are. The total voltage induced in the rotary search coil is given by Figure 5.8:

$$U_s = U_1 \sin v_2 + U_{II} \cos v_2 = U_0(\sin v_1 \sin v_2 + \cos v_1 \cos v_2)$$
$$= U_0 \cos(v_1 - v_2) \tag{5.7}$$

Thus the voltage to the receiver is greatest when the search coil points at the transmitter and least when it points perpendicularly to that direction. Modern direction finders do this automatically, finding the signal minimum by means of a servo system.

One method of achieving this is to modulate the output signal from the loop antenna by a square wave. This shifts the direction of the cardioid by 180° each time the square wave changes its polarity (Figure 5.9). This gives an amplitude modulation to the received signal, and the depth of modulation depends on the direction to the transmitter. When this direction is perpendicular to the plane of the

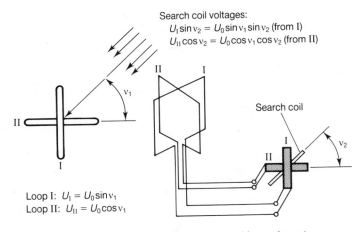

Search coil voltages:
$U_I \sin v_2 = U_0 \sin v_1 \sin v_2$ (from I)
$U_{II} \cos v_2 = U_0 \cos v_1 \cos v_2$ (from II)

Search coil

Loop I: $U_I = U_0 \sin v_1$
Loop II: $U_{II} = U_0 \cos v_1$

Figure 5.8 Crossed-loop antenna with goniometer

loop antenna (CD in Figure 5.9), the output signal is constant (i.e. the depth of modulation is zero), independent of the polarity of the square wave. If this output signal is delivered instead by the search coil, it is zero when the coil has the desired direction. In both cases, the servo system adjusts to zero depth of modulation. There is an ambiguity, however, but it can be solved by means of the phase of the output signal, referenced to the phase of the modulating signal. Modulation frequencies in these cases are of the order of 100 Hz.

Although still in existence, the goniometer as described above is used less today. It has been replaced to a large extent by an all-electronic version, the Automatic Direction Finder (ADF). The ADF also uses crossed-loop fixed antennae (Figure 5.8). For each loop the signal magnitude is measured, and a large number of measurements is averaged (in order to minimize errors). As the directional sensitivity of a loop antenna follows a sine function (Equation (5.1) and Figure 5.6), the ratio between the outputs from two perpendicular loops is (Figure 5.8)

$$U_I/U_{II} = \frac{U_{01}}{U_{02}} \tan v \qquad (5.8)$$

Assuming $U_{01} = U_{02}$ (by means of AGC), we obtain

$$v_1 = \arctan \left(\frac{U_1}{U_{II}}\right) + n \cdot \pi \qquad (5.9)$$

where $n = 0$, 1. In order to solve the ambiguity, a non-directional

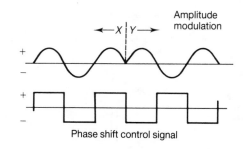

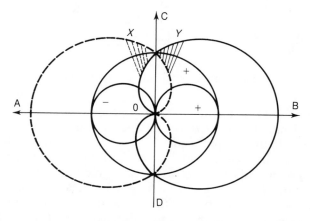

Figure 5.9 Automatic direction finding by modulation of the output signal

sense antenna is used for the ADF. The phase of the sense antenna output signal is compared to the phase of the strongest signal from the two loops. Equal phase means positive direction; opposite phase means negative direction. A block schematic of an all-electronic ADF is shown in Figure 5.10.

Automatic direction finders in one version or the other are standard equipment on board ships and aircraft.

5.4.1 Antennae for improved accuracy

5.4.1.1 The Adcock antenna

As mentioned above, the output voltage of a loop antenna is in proportion to the antenna dimensions (Equation (5.8) below) ex-

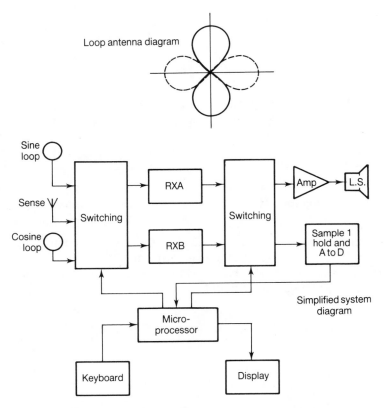

Figure 5.10 ADF antenna pattern and receiver system block diagram (courtesy Standard Marine A/S)

pressed in wavelengths, and for that reason it is desirable to have the largest size antenna possible. At the same time, unwanted signals, e.g. reflections from the ionosphere, should be attenuated to a great extent. One type of antenna meeting these requirements better than the usual loop antenna, is the Adcock antenna. This consists of vertical antenna poles connected to well-shielded, balanced cables (Figure 5.11), and might be regarded as a loop antenna without horizontal parts. The absence of such parts makes the sensitivity to signals reflected from the ionosphere very much lower, as such signals usually have a significant polarization change compared to the direct signal. An Adcock antenna receives vertically polarized electric fields.

Using Figure 5.3 and a point of reference half-way between the two antenna poles making one couple (the same as the vertical sides of a

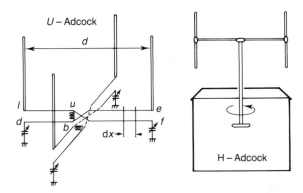

Figure 5.11 The principle of the Adcock antenna

loop antenna), the signal phases in the two antenna poles are $-(2\pi d/2)/\lambda \sin \varphi$ and $(2\pi d/2)/\lambda \sin \varphi$, respectively.

If the antenna height is h, where h $\ll \lambda$, the induced antenna voltage in a homogeneous field is

$$U = E_0 h \left[\exp\left(j\,\frac{\pi d}{\lambda}\sin\varphi\right) - \exp\left(-j\,\frac{\pi d}{\lambda}\sin\varphi\right)\right]$$

$$= E_1 h \sin\left(\frac{\pi d}{\lambda}\sin\varphi\right) \tag{5.10}$$

where E means field strength. (Equation (5.10) applies both to Adcock and loop antennae.) Thus when φ is close to zero, the output voltage is proportional to the angle of incidence. A large directivity consequently implies a large value of d/λ. For medium and short wavelengths the Adcock antenna is usually too large to be mounted on board a ship (20–50 m horizontal distance and 10–20 m high poles, the latter in order to have as great an induced voltage as a loop antenna with many turns), but it is utilized for fixed direction finders on land.

5.4.1.2 Doppler direction finders

A receiver antenna for Doppler direction finding is much larger than one wavelength and is consequently utilized only for VHF and UHF. (Airport Doppler direction finders utilize the communication signals

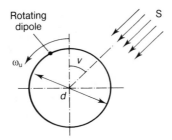

Figure 5.12 The principle of the Doppler direction finder

from the aircraft at 118–137 MHz (civil), and 230–400 MHz (military).) The antenna system consists of a large number of element antennae (of the order of thirty), usually dipoles, mounted on the periphery of a circle. The receiver is switched from one element to the other in sequential order, giving the impression that the receiving antenna makes a smooth rotating movement (Figure 5.12). The frequency of the received signal is changed according to the Doppler principle

$$f_d = \frac{v}{\lambda} = \frac{\omega_u d/2}{\lambda} \sin(\omega_u t + v) = \frac{\pi f_u d}{\lambda} \sin(\omega_u t + v) \qquad (5.11)$$

The phase of this frequency modulation is unambiguously determined by the virtual rotation movement of the antenna and is given by the direction to the transmitter. Demodulation gives a low-frequency signal whose phase varies proportionally to $\cos(\omega_u t + v)$. The amplitude of the phase variation is unimportant as only the frequency maximum or minimum with regard to the rotation is measured. This makes a Doppler direction finder less sensitive to interference than conventional direction finders. The most important reason for using Doppler direction finders is, however, the considerably improved accuracy which follows from the large-dimension antenna making the system less sensitive to reflected signals.

In a practical system there may be problems in ensuring that all element antennae and the corresponding signal channels are identical. Possible differences give phase variations in the received signal, implying errors in the measured Doppler phase. A solution to this is to utilize two counter-rotating movements, taking the difference between these such that the errors cancel (Figure 5.13) (Rohde and Schwarz, 1980).

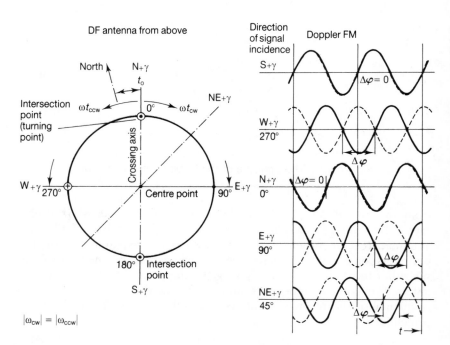

Figure 5.13 The principle of two counter-rotating RF probes. The measured direction of incidence of the received signal is $\Delta\varphi/2$ (courtesy of Rohde and Schwarz)

5.5 Accuracies

The accuracies of different direction finders is strongly dependent on local conditions, such as time of year, time of day (because of ionospheric reflections at MF), distance to the transmitter, receiver quality, etc. The most important local conditions are the siting of the antenna with regard to reflecting objects, for example different parts of a ship, water surfaces, ground, etc. (Gething, 1978). Mounting the antenna high above the environment is often advantageous, but the small (in wavelengths) antennae are not very directional, and it may be difficult to avoid interfering reflections that are independent of the mounting. Usually, the accuracy of a medium wave direction finder is 1–10°, and the coverage varies quite considerably. A Doppler direction finder, on the other hand, whose dimensions are several wavelengths (which would introduce ambiguities in a conventional direction finder because the phase difference between the edges would be larger than 2π), may have an accuracy of 0.1–0.5° (Rohde and Schwarz, 1980).

The advantages of conventional direction finders and the reason for their widespread use is the cheap equipment and the existence of a great number of transmitters.

Advanced signal processing can reduce the influence of many of the error sources which imply low direction finding accuracy. In addition to equipment deficiencies, this applies essentially to signal propagation factors, i.e. reflections from the environment and from the ionosphere. As an example, the use of adaptive pattern techniques for array antennae can be mentioned, where, first, a coarse measurement of the incidence direction of the signal is made, with equal weight on all element antenna signals. Then the output signal from each element (amplitude as well as phase), is weighted such that the receiver, by means of an algorithm, makes an artificial antenna pattern having a narrow main lobe in the measured direction. In this way reflected signals from other directions have reduced influence in accordance with the sidelobe levels in those directions.

⑥ Aircraft systems

6.1 Introduction

Aircraft navigation aids are still developing rapidly, the main reason for this being the rapid development of air traffic itself. With increasing sizes and velocities of aircraft come stricter requirements with regard to traffic flow, safety and independence of weather conditions. The electronics development has made it possible to satisfy these requirements, at costs which are usually acceptable. This has meant improvements of existing navigational aids, and development of new aids. Lack of coordination of solutions has, however, implied an abundance of systems which, in many cases, have similar tasks, so that a rationalization should be possible. The problem is that it is often very difficult, not to say impossible, to agree on which systems should be excluded. National and industrial considerations must be taken into account, and it is desirable to be able to fully utilize investments already made. As far as new systems are concerned, it is especially difficult to determine if one system is technically or economically better than the other. Some of the consequences of all this appear on the exterior of a modern passenger aircraft, where there are a great number of antennae (Figure 6.1).

Some of the systems belonging to these antennae are dealt with in the following pages. Attention is concentrated on civil systems where international standardization has progressed far. (The sub-organization of the United Nations, ICAO (International Civil Aviation Organization) is responsible for standardization issues in international civil aviation. The organization publishes detailed specifications and rules for, among other things, electronic navigation aids.) Only

antenna locations

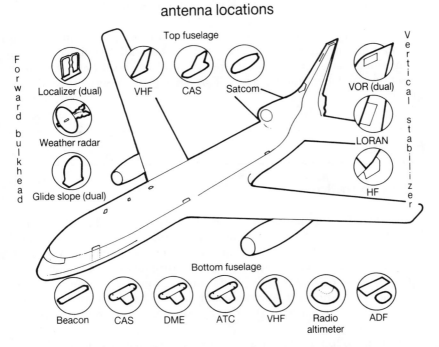

Figure 6.1 Antennae on a modern passenger aircraft

systems exclusively utilized in aviation are dealt with.

This chapter only describes the functions and properties of the systems themselves, but it should be briefly mentioned that combined use of outputs from various navigational aids is very important in aircraft navigation. Such integration of systems and optimized output of processed information is used in so-called Flight Management Systems (FMS), often for the purpose of Area Navigation (RNAV). RNAV permits guidance of an aircraft along any desired course within the coverage of signals from radionavigation stations or within the performance limits of self-contained systems. In FMS, inputs from available navigational sensors (e.g. compass, altimeter, VOR, DME, LORAN-C, GPS, radar, inertial systems, etc.) are used in an optimum manner, usually by means of Kalman filtering (AGARD, 1989), together with other cockpit information (e.g. display control, aircraft flight properties, fuel situation, air data, etc.), in order to provide full functional and operational capabilities for navigation, cockpit management and aircraft avionics management. However, a full description of FMS and RNAV is far beyond the scope of this book.

Civil aviation authorities in different countries publish a mass of information on airports in respective countries and associated aircraft navigational aids, including maps of airport positions, approach directions and conditions, as well as signal characteristics and frequencies.

6.2 VOR

6.2.1 Operation

VOR (VHF Omnidirectional Range) operates in the band 112–118 MHz with 50 kHz channel separation and was internationally standardized in 1949. The frequency selection implies that it is practically insensitive to atmospheric disturbances, but range is limited because of line-of-sight signal propagation. Thus the transmitters are often sited on mountains or hills. The VOR indicates the direction from the aircraft to the transmitter. Transmitter power is about 200 W.

The conventional VOR transmitter antenna has a composed radiation pattern consisting of a non-directional and a figure-of-eight pattern in the same manner as described for direction finders (Figures 5.4, 5.6 and 5.7). However, horizontal polarization is used for VOR. The resulting cardioid radiation pattern is rotated by thirty turns per second. In earlier equipment this was done by mechanical rotation of the antennae, but it is now done electronically. The electronic rotation is realized by mounting two antennae with a figure-of-eight pattern perpendicular to each other (Figure 6.2), fed by a carrier which is multiplied by a 30 Hz signal. The phase of the modulation signal is shifted 90° between the two antennae. A receiver in direction α gets a direction-dependent signal via the two figure-of-eight diagrams plus a direction-independent signal via the circular pattern

$$v_r(t) = \cos \omega_c t \ \ + a \cos \omega_c t \cos \omega_m t \cos \alpha + a \cos \omega_c t \sin \omega_m t \sin \alpha$$

from the non-directional antenna	from one figure-of-eight pattern	from the other figure-of-eight pattern

$$(6.1)$$

This can be rewritten as

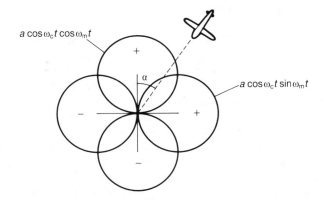

Figure 6.2 The directional part of a VOR transmitter antenna

$$v_r(t) = \cos \omega_c t[1 + a \cos(\omega_m t - \alpha)] \tag{6.2}$$

where ω_c is the angular carrier frequency (at the VHF band), $\omega_m/2\pi = 30$ Hz, and $a \approx 0.3$ (according to standardization require-ments). Equation (6.2) shows that the received signal has constant magnitude when $\omega_m t - \alpha$ is constant, i.e. when $d\alpha/dt = \omega_m$. Thus, Equation (6.2) describes a lobe with a constant rotational velocity, so the phase of the received signal after amplitude demodulation is a linear function of the direction α (Figure 6.2).

Another signal is also transmitted by means of the non-directional part of the antenna. This is amplitude modulated by a subcarrier at 9960 Hz which, in turn, is frequency modulated by 30 Hz. The modulation index is 16 (i.e. ± 480 Hz frequency deviation). Moreover, the signal is also amplitude modulated by a Morse code, a 1020 Hz tone for identification.

Thus the complete signal received from a VOR-transmitter is

$$v(t) = \cos \omega_c t[1 + a \cos(\omega_m t - \alpha)$$
$$+ b \cos(\omega_u t + \beta \cos \omega_m t) + f(t) \cos \omega_i t] \tag{6.3}$$

where $b = 0.3$, $\omega_u/2\pi = 9960$ Hz, $\beta = 16$, $\omega_i/2\pi = 1020$ Hz and $f(t)$ is the Morse tone (depth of modulation about 0.2).

6.2.2 The receiver

The phase of the frequency modulation has been selected so that the modulating signal is in phase with the 30 Hz rotation at the very moment when the cardioid points directly to the magnetic north ($\alpha = 0$) in Figure 6.2. A measurement of the phase difference between the two demodulated 30 Hz signals gives an unambiguous direction. A block diagram of a VOR receiver is shown in Figure 6.3.

The course selector connected to the receiver phase detector is usually set (by the pilot) to the desired radial (i.e. value of α in Figure 6.2) from the VOR transmitter which is to be used for the navigation. This selection implies that the phase difference between the demodulated AM and FM signals is chosen to have a fixed value. A pointer in the cabin shows the deviation from the desired course, i.e. the difference between the predetermined and measured phases. All airliners have VOR receivers, and these are also relatively common in general aviation.

A modern VOR receiver contributes less than 1° to the total error, which is normally between 1 and 5°. The receiver error is mainly path differences of the two 30 Hz signals and phase detector inaccuracies. Most of the total error stems from signal reflections close to the transmitter. This is usually sited as high as possible above the surrounding terrain and buildings, but reflection errors still dominate.

6.2.3 Doppler VOR

One method of reducing the reflection error is to increase antenna size. An FM-signal is also less sensitive to reflections than an

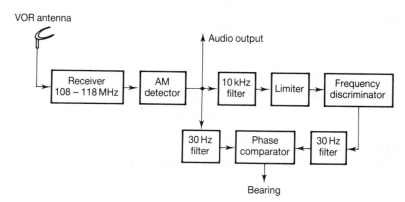

Figure 6.3 Block diagram of a conventional VOR receiver (reprinted from W. R. Fried and M. Kayton (eds): *Avionics Navigation Systems* by permission of John Wiley & Sons, Inc. All Rights Reserved)

AM-signal. These factors are utilized in Doppler VOR. Here a 30 Hz amplitude modulated signal is transmitted by the reference antenna (i.e. the non-directional diagram) instead. The antenna arrangement also consists of a large number (40–50) of element antennae mounted in a ring around a reference antenna. The transmitter signal, amplitude modulated by 9960 Hz, is switched to these element antennae sequentially, such that the signal source virtually makes a circular movement. It must be possible to use the same receiver in a Doppler VOR as in a conventional one, so the rotation frequency of the signal source is 30 Hz. The frequency deviation in the receiver caused by the movement of the signal source must be ±480 Hz (Section 6.2.1), which determines the diameter of the circular antenna arrangement.

The (virtual) velocity of the transmitter is $\omega_m d_a/2$, and the maximum Doppler frequency is then $\omega_m d_a/2\lambda_c$, where d_a is the diameter of the antenna circle, and λ_c is the carrier wavelength. As the maximum Doppler frequency is the received frequency deviation which should be 480 Hz, the antenna diameter becomes

$$d_a = 2 \cdot \frac{480\lambda_c}{\omega_m} \approx 13.3 \text{ m} \tag{6.4}$$

The received signal from a Doppler VOR transmitter thus consists of an amplitude-modulated reference and a frequency-modulated angular signal (FM on the subcarrier 9960 Hz), where the modulating 30 Hz signal is in phase with the reference amplitude modulation when the receiver is exactly to the north of the transmitter. Figure 6.4 shows the received frequency components.

As a conventional VOR transmitter antenna has a diameter of

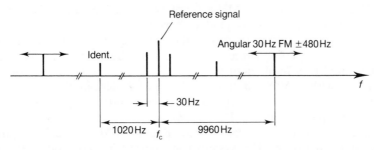

Figure 6.4 Received frequency components from a Doppler VOR transmitter. In a conventional transmitter, on the other hand, the amplitude modulation of the carrier has a directionally dependent phase, whereas the phase of the frequency modulation on the subcarrier is the directionally independent reference

0.5–2.5 m, Doppler VOR means a considerable increase in size (in addition, the counterpoise (i.e. the earth plane) is several times that size). There is also a considerable decrease of the reflection error, of the order of 10 times. The decrease is greater when the points of reflection are situated far away from the line-of-sight between transmitter and receiver, as the phases of the reflected signals then vary more (compared to the phase of the direct signal) when the signal source moves along the circle.

6.3 DME

6.3.1 Principles of operation

DME (Distance Measuring Equipment) is used to measure the distance between the aircraft and the ground station. The radar principle is used, i.e. measurement of the transit time of a pulse from the aircraft to the ground station and back. Carrier frequencies are in the range 962–1213 MHz, and the transmitted pulse power is between 50 and 1000 W. The maximum range is about 370 km, but at a flight level of 3000–6000 m it is about 120 km (only line-of-sight propagation at these frequencies). The system was standardized internationally in 1959.

The DME frequency range is subdivided into 126 interrogation and 126 reply channels with a channel separation of 1 MHz. The interrogation channels have been allocated to the frequency range 1025–1150 MHz; the reply channels to two frequency ranges, 962–1024 and 1151–1213 MHz (applies to the X-channels). Each interrogation channel is coupled to a specific reply channel 63 MHz above or below, depending on the channel used. Each ground transponder has a fixed frequency, whereas the aircraft interrogator can be tuned to a number of frequencies. When the ground station is co-located with a VOR station, the resulting combination forms the standard ICAO $\rho - \theta$ navigation system and is operated on a standard frequency pairing (ICAO, 1985).

The DME pulses are transmitted in pairs with a specified pulse shape and separation. The system can operate in two modes, the X- and the Y-mode. The pulse separation is 12 μs for the X-mode (both interrogation and reply), and 36 μs (interrogation), and 30 μs (reply), for the Y-mode. The Y-mode which uses the same frequency range (1025–1150 MHz) for interrogations as well as replies is not used very much and can be regarded as a capacity reserve. The pulse shape is the same in both and should be approximately Gaussian. It is shown in Figure 6.5.

When the ground transponder receives interrogation pulses from an aircraft, the pulses are detected, regenerated and retransmitted after a specific time such that the total time delay of the transponder is 50 μs (Figure 6.6).

In the aircraft the total time delay between the transmitted and received pulses is measured, and after subtraction of 50 μs from the measured time the distance can be calculated.

6.3.2 Search procedure

As each ground transponder has a fixed carrier frequency, a variation of the pulse pair repetition frequency (PRF) is the only way for an aircraft to recognize its own pulses. Thus this frequency is varied randomly within an allocated range which is 120–150 pulse pairs per second in the search mode, and 24–30 pulse pairs per second in the track mode. When an aircraft starts transmitting interrogation pulses,

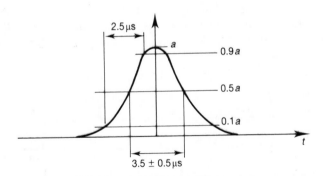

Figure 6.5 DME pulse shape

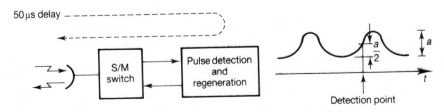

Figure 6.6 Principle of operation of the ground transponder

having entered the range of the ground transponder, it does so with a random PRF which is jittered within 120–150 Hz. A certain time delay, τ, after the transmission of each pulse pair from the aircraft, a time gate is opened in the aircraft receiver. This delay is increased continuously from zero and upwards after the search procedure has begun. The time gate is open for 20 μs each time, which corresponds to 3 km in space (remember that pulses travel back and forth), and it moves away from the pulse transmission instant with such a velocity that it spends 20 s searching for reply pulses from a transponder distance of 0–370 km (this velocity corresponds to 18 km/s in space). In other words, the time delay τ increases linearly with time as $\tau = 18t/150$ (τ in ms, t in s). At an average PRF of 135 Hz the time gate is open $20 \times 135\ \mu$s/s = 2.7 ms/s.

The ground transponder transmits 2700 pulse pairs per second in total, irrespective of the interrogation rate (Section 6.3.3). So, on an average, the aircraft receiver receives between seven and eight pulse pairs per second ($= 2700 \times 0.0027$) of all that is transmitted by the transponder (Figure 6.7). The opening of the time gate is synchronized to the transmission of the interrogation pulses of that aircraft (time delay τ above). As the time gate opening corresponds to 3 km in space, and as it moves at 18 km/s in space, it takes 1/6 s to pass a certain point in space. During that time, the ground transponder replies to the aircraft's interrogations at the same rate as these are received, i.e. at 135 pulse pairs/s. Thus the aircraft receives

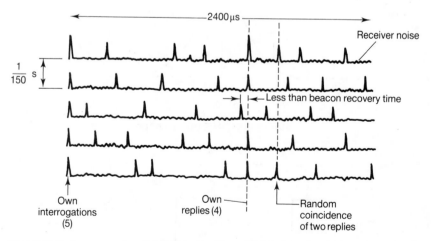

Figure 6.7 An example of pulses entering a DME aircraft receiver (reprinted from W. R. Fried and M. Kayton (eds): *Avionics Navigation Systems* by permission of John Wiley & Sons, Inc. All Rights Reserved)

$135/6 = 22.5$ pulse pairs while the time gate passes the distance corresponding to the transponder distance. As other received reply pulses belong to other aircraft with other PRFs, the opening of the time gate is not synchronized to the time of arrival of these (Figure 6.7). The receiver counts the received and detected reply pulses and obtains a result which is, on average, about three times (9 dB) larger when the time gate passes the position corresponding to the reception of the replies to the transmitted interrogation pulses from that aircraft, than otherwise. When this happens, the search is stopped, and the time gate is centred around the point which gives the highest number of detected reply pulses. The receiver switches to the tracking mode, transmitting 24–30 pulse pairs/s. As the aircraft approaches the ground transponder, the time gate tracks the aircraft movement in such a way that it is always centred around that maximum point.

6.3.3 The transponder

A DME ground transponder is designed to serve a maximum 100 aircraft simultaneously, with a distribution of 95 in the tracking and five in the search mode. A transponder transmits an average $\eta(95 \times 27 + 5 \times 135)$ pulse pairs/s, where η is the reply efficiency of the transponder, usually about 80 per cent; $\eta = 0.8$ gives 2592 pulse pairs/s. However, there are hardly ever so many aircraft within 370 km (200 n.m.). Not even at the busiest airports. Previously, it was desirable that the AGC of the aircraft receiver had an approximately constant PRF. For this (now historical) reason, noise pulses are used for triggering the transponder until the transmitted number of reply pulses, plus regenerated noise pulses, is at 2700 ± 90 s. η is not 100 per cent because there is a dead time in the transponder for (usually) 60 μs after the reception of an interrogation, and pulse pairs from other aircraft arriving in this time interval do not trigger any reply. The task of the transponder dead time is to prevent echo pulses, i.e. reflections from the environment, from being regenerated. Another reduction of the reply efficiency is caused by the identification transmitted about every 30 s and consisting of regular pulse pairs with PRF = 1350 Hz keyed by a three-letter Morse code (1/8 s per dot and 3/8 s per dash).

When the aircraft equipment does not receive a reply to its interrogations, a memory is used such that the time gate either stays at the position last measured, or, more commonly, moves at the velocity corresponding to the rate of change of the last measured distance. In this way extrapolations of the measured distance can be made up to about 10 s.

6.3.4 Accuracies/Use

The accuracy of a DME system is usually in the range 100–300 m. The largest source of error is the inaccuracy of the equipment itself, especially the inaccuracy of the 50 μs time delay though the transponder (ICAO, 1985), accepts ± 1 μs of variation), and the inaccuracy of the half-amplitude detection (Figure 6.6). Another important source of error is signal reflection from the environment, implying a distortion of the received pulses and thus also a detection error.

DME is a navigational aid used in terminal areas as well as between airports. When two or more DME transponders are established to cover a large contiguous area, the system can be used as an independent navigation system for so-called ρ–ρ navigation.

6.3.5 DME/P

DME is used, together with the new microwave landing system (MLS, see following section), for distance measurement, thus giving the third coordinate of the landing aircraft. Conventional DME (also called DME/N) is too inaccurate for such use. A more accurate version is therefore standardized by ICAO. The principal operation is the same, but steeper pulses are used, and consequently larger bandwidth, as well as another type of detector in the receiver. In addition, the number of channels is increased to 200 in order to meet the requirements of MLS.

A pulse shape satisfying ICAO requirement is achieved if the rising edge of the pulse follows a cosine curve and the declining edge a squared cosine (Kelly, 1984). The nominal rise time (10–90%) should not be more than 1600 ns. Between the levels $0.05a$ and $0.30a$, where a is the pulse amplitude, the rise time should be 250 ± 50 ns with a linearly rising edge. The rise time is not allowed to vary more than 20 per cent from the linear average between the two levels. The detailed shape of the declining edge is left to the designer, but the whole pulse should have a length between the 50 per cent levels as in the DME/N, i.e. 3.5 μs nominal value. The pulse is shown in Figure 6.8.

The reasons for this pulse shape are closely connected to the

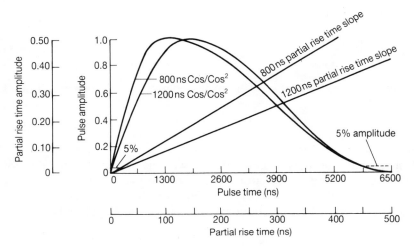

Figure 6.8 DME/P pulse shape

requirement that the signal reflections should not (or very little) influence the accuracy and to the use of a special type of detector. The latter, called DAC (Delay, Attenuate and Compare), works according to Figure 6.9. A delayed version of the pulse is compared to an attenuated version of the same pulse. The two rising edges should cross each other in the linear area, i.e. between 5 per cent and 25 per cent of the amplitude. If the gain is A ($A < 1$), the linear rise

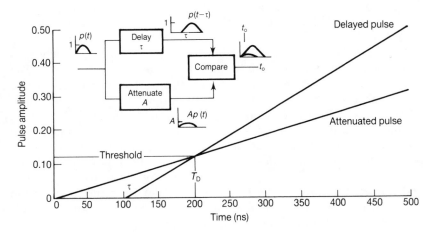

Figure 6.9 DME/P detection principle

time T_r, and the time delay τ, we have

$$A \frac{t}{T_r} = \frac{t - \tau}{T_r} \tag{6.5}$$

which gives the time of crossing of the edges, i.e. the detection time

$$t_d = \frac{\tau}{1 - A} \tag{6.6}$$

Notice that the pulse amplitude as well as the rise time do not influence the time of detection because of the linear slope.

Equations (6.5) and (6.6) give relative threshold values of the detection points, both for the delayed and the undelayed pulses:

$$\frac{t_d}{T_r} = \frac{\tau}{(1 - A)T_r} \tag{6.7}$$

$$\frac{t_d - \tau}{T_r} = \frac{A\tau}{(1 - A)T_r} \tag{6.8}$$

A common value of the delay time is 100 ns. If the gain A is chosen to be 0.5 (i.e. -6 dB) and the rise time $T_r = 800$ ns, the detection levels of the attenuated and the inattenuated and delayed pulse are $0.25a$ and $0.125a$, respectively, i.e. -12 and -18 dB, respectively, related to the pulse amplitude (Figure 6.9). This low detection point (i.e. detection after only 200 ns of the attenuated pulse), reduces the influence of reflections drastically, compared to the case of DME/N. Most reflections simply do not arrive in time (cf. LORAN-C).

It should be emphasized further that the point of detection is independent of the amplitude and rise time of the pulse. The transponder delay is the time between the start of the received and the transmitted pulses, called virtual origin by ICAO in their Standards and Recommended Practices (SARP) (1972). It is defined as the point on the time axis which is intersected by a straight line through the 5 per cent and 30 per cent values of the rising edge of the pulse. The transponder delay is kept within the tolerance limits by means of pilot pulses, i.e. pulses generated by the transponder itself and transmitted through the entire transponder from the input to the output with a simultaneous measurement of the transit time. In this way, received and transmitted pulses, as well as pilot pulses, can have different amplitudes and rise times without influencing transponder delay. The given tolerances ±20 per cent of non-linearity (between 5 per cent and 30 per cent), $\tau = 100$ ns ± 10 ns, and $A = -6$ dB ± 1 dB with such a detector mean less than 3 m of distance error.

The steeper pulse (compared to DME/N) implies that the frequency spectrum is wider. ICAO, however, has retained its requirement with regard to power leakage into the adjacent channels such that the maximum levels allowed into adjacent channels are the same (in watts) for DME/N and DME/P. Consequently, the transmitted power of DME/P must be lower (as the spectrum is wider), which means a reduced range. (The transmitter power from the aircraft is about 250 W and from the ground transponder about 300 W.) This problem has been solved in the following way.

The highest accuracies are required only in the neighbourhood of the airport. Thus DME/N is utilized for the approach, until about 8 n.m. from the runway. A transition zone goes from 8 to 7 n.m. and, at distances below 7 n.m., DME/P is used by the aircraft (Figure 6.10). The two modes are called initial approach (IA) and final approach (FA).

DME/P IA is, therefore, exactly the same as DME/N except for the interrogation frequency (see below), and the range (22 n.m.). The aircraft interrogator changes during the transition zone from the IA to the FA mode, i.e. from 12 to 18 μs of pulse separation in the pair,

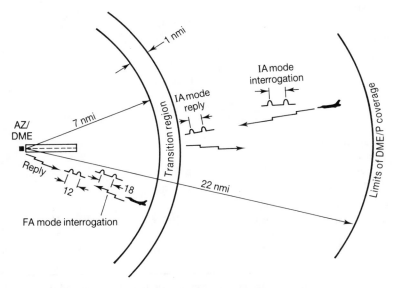

Figure 6.10 Two DME modes are used during approach

and simultaneously switches the ground transponder from IA to FA, i.e. from DME/N to DME/P. In this way a DME/P transponder is capable of serving aircraft with DME/N as well as DME/P. The interrogation frequency is 40 Hz in the IA search and FA modes, and 16 Hz in the IA tracking mode, but the receiver bandwidths are 300–400 kHz in the IA mode and 3.5–4.5 MHz in the FA. The latter bandwidth is a prerequisite to be able to utilize the steep pulses.

The accuracy requirements have implied that the SARP of ICAO recommends two standards. Standard 1 must satisfy the requirements for CTL (Conventional Take-off and Landing), where a radar altimeter is used by the aircraft during the very last phase of the landing. Standard 2 applies to STOL and VTOL (Short Take-off and Landing and Vertical Take-off and Landing), as well as all landings where MLS is used on descents to the ground. Table 6.1 shows the error

Table 6.1 Distance error tolerances (95%) for DME/P in the IA and FA modes for the two standards

Location	Standards	PFE	CMN
20 min to 5 min from MLS approach reference datum	(1) and (2)	± 250 m reducing linearly to ± 85 m	68 m reducing linearly to 34 m
5 nmi to MLS approach reference datum	(1)	± 85 m reducing linearly to ± 30 m	18 m
	(2)	± 85 m reducing linearly to ± 12 m	
	IA[a] mode	± 100 m	34 m
At MLS approach reference datum and through runway coverage	(1)	± 30 m	18 m
	(2)	± 12 m	12 m
Throughout back azimuth coverage volume	(1) and (2)	± 100 m	68 m
	IA[a] mode	± 100 m	68 m

Note: [a]At distances from 5 nmi to the MLS approach reference datum and throughout the back azimuth coverage the IA mode may be used when the FA is not operating.

tolerances in both cases (95 per cent). PFE (Path Following Error) is the physical deviation of the aircraft from the measured distance, filtered through a lowpass filter with a cut-off frequency of 0.5 rad/s. CMN (Control Motion Noise) are errors in the aircraft control loops caused by the distance measurement, filtered through a bandpass filter with cut-off frequencies 0.3 and 10 rad/s.

As no new frequencies have been allocated to DME/P compared to DME/N, the extra channel capacity has to be realized by means of variation of the pulse intervals in the pairs. This has already been described above in connection with the transition by the aircraft between the IA and FA modes. There are also other pulse intervals than those mentioned (namely $36 \rightarrow 42\ \mu s$, $24 \rightarrow 30$ and $21 \rightarrow 27$, called Y-, W- and Z-channels, respectively), such that the total channel number is 200. In all the channels the interrogator increases the pulse interval by $6\ \mu s$ during transition from the IA to the FA mode, whereas the transponder uses the same interval in both modes.

6.4 ILS

6.4.1 System principles

ILS (Instrument Landing System) resulted from developments in Germany and the United States since the 1920s. It has been in civil use as standardized by ICAO since 1947 and is now in common use all over the world.

The ILS ground equipment consists of three main parts transmitting information to the aircraft (Figure 6.11): one localizer giving azimuth information and sited at the far end of the runway, one glide path transmitter giving elevation information and sited offside the runway touch-down point, and three or less markers giving distance information as the aircraft passes. The markers are sited about 8 km, 1 km and 400 m from the touch-down point (outer, middle and inner marker). Many airports have only the outer and middle markers.

ICAO defines three visibility ranges for instrument landing of civil aircraft (1985): category I means 60 m ceiling and 550 m ground visibility; category II means 30 m ceiling (or decision height), and 350 m visibility; and category III means 0 ceiling and 200 m (III a), 50 m (III b) and 0 m (III c) visibility. ILS ground equipment belonging to one of these categories gives useful signals down to an altitude of 60, 15 and 0 m, respectively. Most ground equipment

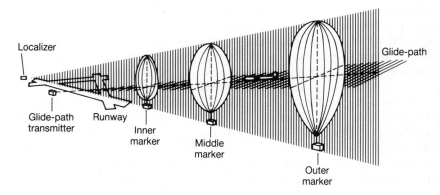

Figure 6.11 ILS

belongs to category I, but the amount of category III equipments is increasing. (It is worth noting that the categorization of an airport also means requirements that do not concern the landing system, e.g. runway width, illumination, etc. Thus there may be category I airports with category II or III ILS equipment.)

6.4.2 Markers

A marker consists of a transmitter at 75 MHz and an antenna transmitting a fan lobe upwards across the approach direction. (Figure 6.11). The outer marker transmits a signal modulated by a 400 Hz tone, two dashes per second; and the signal from the middle marker is modulated by 1300 Hz, dash–dot every $\frac{2}{3}$ second (the inner marker: 3000 Hz, 6 dots/s). In addition to the signal in the headset of the pilot when passing the marker, there is a cockpit visual signal coloured to denote which marker is being passed (outer marker purple, middle yellow, inner white). The distance from the runway to the markers is such that the middle marker is sited where the aircraft passes the decision height of category I (60 m), and the inner where the aircraft passes this altitude for category II. The outer marker is nominally sited 3.9 n.m. from the threshold but if unpractical, between 3.5 and 6 n.m. The middle marker is situated about 1050 m from the beginning of the runway (the threshold). The inner marker is used very little, but if there is one it should be situated 75–450 m from the threshold. The markers transmit wide lobes, about ±40° in the flight direction and about 170° across.

6.4.3 Angular information

The localizer transmits at a carrier frequency in the frequency range 108–112 MHz (50 kHz channel bandwidth), radiating, in principle, two overlapping lobes in somewhat different directions (Figure 6.12). The lobe radiated to the right, as seen from the antenna, contains a carrier signal amplitude modulated by 90 Hz. The other lobe contains the same carrier signal, but modulated by 150 Hz. To the right of the plane where the lobes intersect, the modulation depth of the 90 Hz signal is larger than that of the 150 Hz signal, but to the left of the intersection plane the 150 Hz signal dominates. In the plane of intersection the modulation depths of the two low-frequency signals are equal. The difference in depth of modulation thus becomes an expression of the position of the aircraft related to the desired approach direction and is denoted DDM (Difference in Depth of Modulation). On board the aircraft there is a receiver which is coupled to a cross-pointer showing DDM (Figure 6.13).

The signal from the glide-path transmitter (a typical transmitter power is 20 W) has a carrier frequency in the range 329–335 MHz but does not differ in any other way from the localizer signal, and 90 Hz dominates above and 150 Hz below the plane of intersection between the two lobes. This plane, the glide-path plane, makes an angle of about 2.5–3.5° to the ground, thus forming the nominal glide slope descent angle of the aircraft.

The cross-pointer (Figure 6.13), has, as indicated by the name, two perpendicular pointers which indicate if the aircraft is on the correct course. If the aircraft is to the right of the course line, i.e. the pilot should turn to the left, the perpendicular pointer moves to the left; if

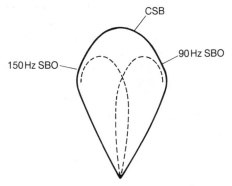

Figure 6.12 Simplified localizer radition diagram. SBO = sideband only, CSB = carrier + sideband

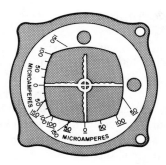

Figure 6.13 Cross-pointer

the aircraft is too low, i.e. the pilot should fly up, the horizontal pointer moves upwards on the instrument.

The operational principle used for transmitting the correct angular information from an ILS ground equipment can be illustrated by considering a system of three equal, non-directional antennae connected to a distribution network, as shown in Figure 6.14. The carrier signal is fed to a power divider and then on to two modulators where it is modulated 40 per cent by 90 and 150 Hz respectively (both modulation frequencies are usually generated by multiplication from the same 30 Hz source). After that, the two signals are fed to a hybrid (the antenna bridge), where the sum output gives a carrier 20 per cent modulated by both frequencies (carrier + sidebands, CSB), whereas the difference output gives sidebands only (SBO). The signal which contains sidebands only is divided and phase shifted by means of different feeder line lengths such that the phases of the signals for the two outer antenna paths are 90° (at the carrier frequency) at each side of the centre antenna carrier phase and with the 90 Hz and 150 Hz sidebands out of phase, as shown by Figure 6.14. (A real transmitter antenna contains many more element antennae, but the subdivision is principally the same.)

A receiver at an angular position φ referred to the symmetry line of the element antennae then gets a signal which can be written

$$v(t) = (1 + 0.2 \cos \omega_{90} t + 0.2 \cos \omega_{150} t) \cos \omega_c t$$
$$+ 0.1(\cos \omega_{150} t - \cos \omega_{90} t) \sin (\omega_c t + \alpha)$$
$$+ 0.1(\cos \omega_{90} t - \cos \omega_{150} t) \sin (\omega_c t - \alpha) \qquad (6.9)$$

where the signal from the centre element is used as a reference and $\alpha = (2\pi/\lambda_c) d \sin \varphi$. The expression can be simplified into

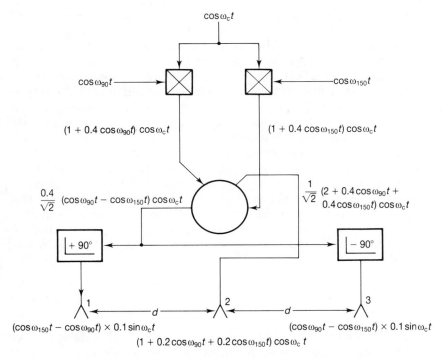

Figure 6.14 Principle for feeding the ILS transmitting antennae

$$v(t) = [1 + 0.2(1 - \sin \alpha) \cos \omega_{90} t$$
$$+ 0.2(1 + \sin \alpha) \cos \omega_{150} t] \cos \omega_c t \qquad (6.10)$$

The depths of modulation of the two low-frequency signals are thus $0.2(1 + \sin \alpha)$ and $0.2(1 - \sin \alpha)$, respectively, and the difference is

$$DDM = 0.4 \sin \alpha \qquad (6.11)$$

Close to the symmetry line of the antennae, φ, and consequently α as well, is small such that

$$DDM = 0.4 \sin \left(\frac{2\pi d}{\lambda_c} \sin \varphi \right) \approx \frac{0.8\pi d}{\lambda_c} \varphi \qquad (6.12)$$

A glide-path antenna operating in the way described by Figure 6.14 is called a *null reference* because the SBO part gives a zero signal along the course line. A real glide-path antenna is usually designed as two element antennae, the upper at about 9.55λ ($\approx$9 m) above the ground and radiating the SBO signal, and the lower at half the height of the former and radiating CSB. Assuming that the ground is a perfect conductor, the signal in space consists of the vector sum of the direct and reflected signals, the latter being virtually radiated from image antennae below ground level, and 180° out of phase with the former (because of the reflection of the horizontally polarized signal). In this way both the CSB and SBO signals are zero on the ground (0° elevation). The CSB radiation pattern has its higher nulls at 2θ, 4θ, etc., where θ is the glide-path angle. The SBO pattern has its nulls at θ, 2θ, etc. The disadvantage of this arrangement is, however, a relatively high sensitivity (by the SBO antenna) to signal scattering from terrain irregularities in front of the antenna (see below). This can be considerably reduced by using more element antennae (Young, 1980). The approach course should be $\alpha = 0$. But as $\alpha = (2\pi d/\lambda_c)\sin\varphi$, there always exist several courses having $\alpha = 0$, i.e. all φ satisfying

$$\frac{2d}{\lambda_c}\sin\varphi_n = n \tag{6.13}$$

where n is an integer. The direction of these so-called false courses depends on the value of d, but ICAO permits none within ±35° (this applies to the localizer, as false glide-path courses do not of course create any danger).

6.4.4 Accuracy/Sources of error

The localizer accuracy is 0.1–0.5° in a sector of ±2.5°, and the range is about 40 km. The glide-path accuracy is usually 0.1–0.2°, and the range is about 18 km. The angular coverage of the localizer is about ±10° in azimuth within 25 n.m., ±35° within 17 n.m. and 7° in elevation. The glide-path angular coverage is about ±8° in azimuth and $0.45\theta - 1.75\theta$ in elevation, where θ is the nominal glide-path angle.

The methods of angle determination in the aircraft and the error tolerances of the different categories are specified in detail by ICAO (1985). The deviation of the cross-pointer from zero is proportional to the input current, so all angles and angular errors are given in microamperes. The maximum deviation of the instrument when measuring the localizer angle, 150 μA, should correspond to a DDM of 0.155 at an angular error of 2° (as seen from the antenna at the far end of the runway). DDM can be regarded as linear for angles up to 3–4°. Acceptable irregularities, so-called bends, of course line are 5 μA, i.e. DDM about 0.005, within the last kilometer before the runway of a category-II localizer.

The maximum deviation of glide-path measurements is DDM = 0.175 (150 μA), corresponding to an angular deviation of $\theta/5$, where θ is the glide slope angle.

The localizer also transmits backward signals which can be utilized. The pilot must be aware that when this so-called backlobe is used, the cross-pointer moves in the opposite direction. Because of the location, glide-path antennae are not designed with the corresponding backward properties.

Due to the surrounding terrain, buildings, etc., signals with DDM $\neq$ 0 are reflected towards the course line to a smaller or greater extent such that bends arise. In order to counter this the antenna systems are made as directional as possible. This makes them rather bulky, with the antenna system of a localizer often as long as 20–30 m. As long as the ground is even and has constant electric properties, the null-reference antenna is a good solution (Figure 6.15), even if levelling of several hundred metres in front of the antenna is necessary. In the opposite case the result is less successful. Different antenna types also have different properties when it comes to low-elevation angle illumination of terrain or buildings (cf. what has been said above about the null reference antenna) (Young, 1980).

The location of ILS, VOR and DME installations at the airports are often coordinated. Then the frequencies of the different aids are also fixed so that the pilot tunes his receiver only to a certain channel to find the correct frequency of ILS-VOR-DME at the same time.

As shown, ILS has only one course line, which reduces the capacity of the system. In spite of the large antennae there are often reflection problems as well, reducing the performance of the system or making

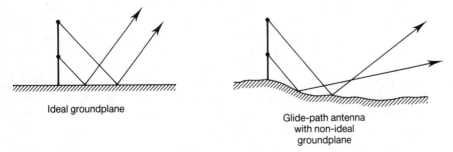

Ideal groundplane

Glide-path antenna
with non-ideal
groundplane

Figure 6.15 Reflections at the glide-path antenna

it unusable at certain airports. Among other things this implies that the distance between approaching aircraft must be relatively large (3–10 n.m.). A further disadvantage of ILS is that it has only forty channels, which may create problems where the airport density is high. (The reason for the small number of channels is that localizer and VOR frequencies are interleaved in a limited frequency range. The glide-path uses a wider frequency range, but as it is linked to the localizer frequencies (ICAO, 1985) the restriction mentioned above applies.) Another problem in certain countries is interference from broadcasting transmitters in the upper part of the FM band. Because of these weaknesses it has been decided by ICAO that ILS shall be gradually replaced by other and better landing aids (see next section) which do not suffer from these problems to the same extent. ILS is protected by ICAO till 1998. (Protection means that operational equipments for international civil aviation must meet the requirements of ICAO.)

6.5 MLS

6.5.1 Introduction

As described in the preceding section, ILS has some principal weaknesses, making it difficult for the system to meet present and future requirements of a landing system. ILS ground equipments have the categories I, II or III and most of them meet the requirements for category I only. The increased regularity and safety which category II or III implies means requirements which may be difficult or impossible to fulfil by the ILS equipment. There are even a few airports where category I performance cannot be achieved by ILS. However, ILS performance has been greatly improved during the last decade.

For these reasons, work was carried out over a period of time to design a landing system which was able to meet these strict requirements. The requirements of this system were worked out by several organizations, beginning in 1967, in the Special Committee 117 (SC 117) of RTCA (Radio Technical Commission for Aeronautics, 1970). The committee was composed of representatives of governments and the aircraft and electronics industry, mainly from the United States, but also from many other countries. Its work was finished in 1970 and the result was presented to the FAA (Federal Aviation Administration), which since then has initiated very extensive investigations and tests to evaluate the system concepts proposed by the committee. Similar work was also performed by ICAO in its AWOP (All Weather Operations Panel, 1978), the result of which has been the basis of an international agreement on system selection.

It was obvious early on that the new system, because of the above-mentioned antenna and reflection problems, would have to operate at microwave frequencies, so the general name was MLS, (Microwave Landing System). Two proposals were evaluated by SC 117: Doppler MLS (DMLS) and the time reference scanning-beam system (TRSB). DMLS had been developed in the United Kingdom, TRSB in Australia and the United States. In 1975 the United States selected TRSB and since then has promoted this system heavily in ICAO as well. A number of trials and investigations were carried out. Finally, after lengthy and sometimes excited discussions, TRSB was also selected by ICAO (1978). It was the opinion of many states, and it still is, that the decision was taken too early, but obviously the decision was required to enable participating states to concentrate on the extensive work necessary to develop the system. The ICAO decision was taken in 1978, but the change from ILS to MLS has been postponed several times. At present, the situation is that this change should be completed by the year 2000.

6.5.2 TRSB angular information

In the time reference scanning-beam system, a so-called 'to–fro' principle is utilized for all angle measurements. The transmitters are located on the ground at approximately the same positions as described previously for ILS, and the carrier frequency is between 5.03 and 5.09 GHz. A fan beam (Figure 6.16), is scanned through the coverage area with a constant scan velocity. In the aircraft a pulse is received every time the beam is scanned. By measuring the time between the passage in one direction ('to' pulse), and the passage in the opposite direction ('fro' pulse), the direction can be determined (Figure 6.17). Using the notation given in the diagram we have

$$t = t_B - t_A = t_2 - t_1 + 2(\varphi_0 - \varphi)/V \qquad (6.14)$$

where

$$V = 2\varphi_0/(t_1 - t_0) \qquad (6.15)$$

i.e. the scan velocity in °/s.

This scan velocity has been standardized by ICAO to $0.02°/\mu s$

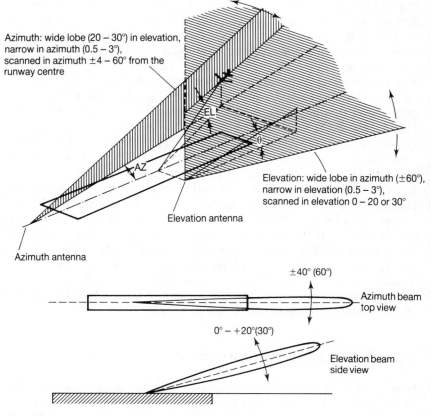

Azimuth: wide lobe (20 – 30°) in elevation, narrow in azimuth (0.5 – 3°), scanned in azimuth ±4 – 60° from the runway centre

EL

AZ

0

Elevation: wide lobe in azimuth (±60°), narrow in elevation (0.5 – 3°), scanned in elevation 0 – 20 or 30°

Elevation antenna

Azimuth antenna

±40° (60°)

Azimuth beam top view

0° – +20°(30°)

Elevation beam side view

Figure 6.16 Principle of operation of the time reference scanning beam system

(1985). ICAO has also given values for $T_0 = (t_B - t_A)_{\varphi=0}$ of the different angular functions. Using the same notation we obtain

$$\varphi = \frac{V}{2} (T_0 - t) \tag{6.16}$$

Table 6.2 gives values of the different angular functions. Here, T_m is the time of an entire scan (i.e. $t_3 - t_0$ in Figure 6.17), and T_v is the length of the break at the turning of the scanning beam (i.e. $t_2 - t_1$ in Figure 6.17). The flare elevation is the final part of the landing when the point of gravity of the aircraft changes its direction of movement from the glide path to an almost horizontal direction at the very

Table 6.2 Values of the MLS angular functions

Function	Scan angle (°)	T_m (µs) for max. scan angle	T_0 (µs)	T_v (µs)	V (°/µs)	No. of scans/s
Approach azimuth	−62 – +62	13000	6800	600	0.02	13 ± 0.5
Approach azimuth, high up-dating rate	−42 – +42	9000	4800	600	0.02	39 ± 1.5
Missed approach azimuth	−42 – +42	9000	4800	600	−0.02	6.6 ± 0.25
Approach elevation	−1.5 – +29.5	3500	3350	400	0.02	39 ± 1.5
Flare	−2 – +10	3200	2800	800	0.01	39 ± 1.5

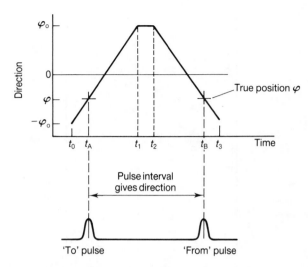

Figure 6.17 Direction of the azimuth beam as a function of time

moment the wheels touch the runway. Only the most sophisticated MLS ground equipments have a subsystem for this. At those airports which do not have such a subsystem and where the requirement for azimuth coverage is ±40° at most, ICAO recommends that the azimuth updating is done at 39 Hz.

The scanning beam principle does not require any directional reference. However, it is required that the scan velocity be constant and known, and that the lobe shape be symmetrical and independent of the direction of the beam. Most principles for detection in the receiver are based on these symmetry properties, either with a fixed (Figure 6.18) or a floating detection threshold. The receiver detects the mid-point between the times where the received pulse exceeds the

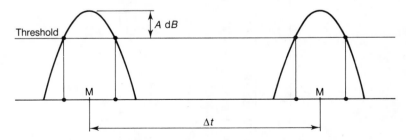

Figure 6.18 Principle of the detection of time-of-arrival of the 'to' and 'fro' pulses in an MLS receiver

threshold on the rising and the declining edge. This mid-point is taken as the time of arrival or the pulse. The detection is preceded by sampling and digitizing of the received pulse and some signal processing to reduce the influence of reflections distorting the pulse shape. The pulse shape reflects the radiation diagram of the ground antenna, including sidelobes and reflections. Even if the transmitter antenna is designed with regard to minimum radiation in unwanted directions, some reflections cannot be avoided, from the terrain as well as from nearby buildings. The resulting pulse distortion implies measurement errors, as indicated by Figure 6.19. The receiver has a time gate operating in approximately the same way as described previously for DME. Its width corresponds approximately to the width of the main lobe of the ground antenna, and when the receiver is tracking, it is centred around the maximum of the received signal. Thus the time gate does not pass the sidelobes, and it also effectively blocks reflected signals which do not coincide with the direction of the main lobe of the signal (if the direct signal is the strongest, which is usually the case). Figure 6.19 shows what happens when the direct and reflected signals partly coincide.

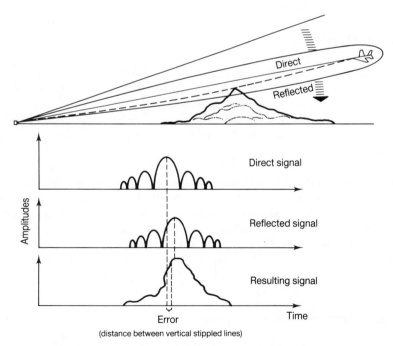

Figure 6.19 An example of signal reflection as a source of pulse distortion

6.5.3 The data message

A data signal is transmitted on a fixed antenna which is co-located with the azimuth antenna to cover the whole scan sector. MLS is a time-shared system utilizing the frequencies 5031.0–5090.7 MHz (ICAO, 1985) allocated to 200 channels (five times as much as ILS), with a separation of 300 kHz. All the angular frequencies and the data signal are therefore transmitted in time-sharing at the same frequency (channel) for each ground installation. The signal format is indicated in Figure 6.20. Every function starts by the transmission of a pure carrier for 768 μs (corresponding to 12 bit-lengths of 64 μs each). After that there is a sequence containing a 5-bit Barker-code as preamble, and function identification of 7 bits. The bit-rate is 15,625 bauds, and the type of modulation is differential phase-shift keying. The signal format is accurately specified by ICAO and contains the angular functions mentioned in Table 6.2, as well as basic data and auxiliary data. The basic data contain different information about the MLS installation in question, including coverage, minimum glide-path angle, DME status, lobe widths and signal quality.

6.5.4 Use

The localizer and glide-path transmitters of MLS have approximately the same location at the airport as for ILS. During a transition period, the two systems have to coexist side by side at some airports, which may imply certain shadowing or reflection problems. Notice that distance is measured by means of DME/P. Thus there is no marker at an MLS airport without ILS.

There is great freedom in the aircraft with regard to the presentation of MLS information. Most designers so far have used the same presentation as for ILS, i.e. a cross-pointer for the angular information. One of the main advantages of MLS is the coverage volume. This is usually utilized in such a way that a required approach course is programmed into the computer on board according to instructions from air traffic control. After that the cross-pointer shows the deviation from the programmed value.

6.5.5 Accuracies and coverage

The angle errors of MLS are usually far less than those of ILS. In the

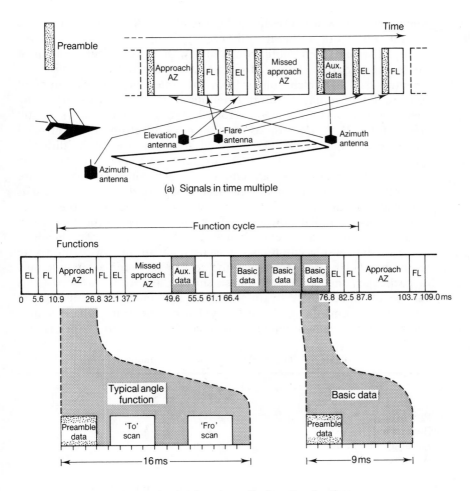

(a) Signals in time multiple

(b) Typical signal cycle and magnification of two functions

Figure 6.20 TRSB signal format

same way as for DME/P requirements are specified as PFE and CMN (Redlien and Kelly, 1978). Measured PFE values are usually in the range of 0.03–0.05° in azimuth, whereas ICAO tolerates 0.18° at the threshold, increasing to 0.36° at 20 n.m. The measured azimuth value of CMN is usually in the range 0.02–0.03°, whereas ICAO tolerates 0.09° at the threshold, increasing to 0.11° at 20 n.m. In elevation the errors are less.

As an example of a real MLS ground installation the data of one commercial MLS are presented in Table 6.3. As shown, there are several versions of TRSB for different types of airports, consisting of more subsystems and subfunctions and with better accuracy than at the larger airports. This is, of course, also reflected by the costs, an essential part of which comes from the antennae. Development of suitable antennae has therefore been one of the main tasks during the work on TRSB (Lopez, 1982; Howell, 1987), and this work is still going on (cf. ILS). Because of the scan rates, electronically controlled antennae are necessary. The selected solutions are either different types of arrays (Howell, 1987), or lens antennae. In the array antennae the phase and amplitude of each antenna element are controlled such that the required beam shape and scan movement is obtained. For the lens antennae, the signals are routed to the feeder elements which radiate towards a lens, or possibly a reflector, which then shapes the final antenna beam. The latter type is the cheapest, but also the least accurate, so it is used at airports with moderate requirements. In order to reduce the cost of the large array antennae in particular, the number of phase shifters has to be reduced. This can be obtained partly by thinning the antennae, i.e. reducing the number of elements, and partly by letting the same phase shifter control several element antennae by a coupling network. Work is also

Table 6.3 Typical MLS ground equipment data

| | (runways up to 2700 m) | |
	AZ	El
Proportional coverage	± 40°	0.9–7.5°
Beam width	3°	2°
Accuracy at threshold	± 6 m	± 0.6 m
Accuracy at 20 n.m.	± 0.29°	± 0.2°

| | (runways up to 3600 m) | |
	AZ	El
Proportional coverage	± 40°	0–20°
Beam widths	2°	1°
Accuracy at threshold	± 4 m	± 0.6 m
Range	20 n.m.	20 n.m.

| | (runways up to 4500 m) | |
	AZ	El
Proportional coverage	± 60°	0–20°
Beam widths	1°	1°
Accuracy at threshold	± 4 m	± 0.6 m
Range	20 n.m.	20 n.m.

going on, in Norway and other countries, on integration of the phase shifter and the power amplifier of the antenna element. The total transmitter power is 10–20 W. Polarization is vertical. Typical antenna dimensions are 1.8 m × 4.8 m for an azimuth antenna with a beam width of 1° (a length of 2.4 m then gives 2° of beam width). An elevation antenna with a beam width of 1° is typically 0.3 m × 4 m. Azimuth antennae often consist of arrays of vertical, slotted waveguides, whereas the typical elevation antenna is a linear array of dipoles.

6.5.6 Costs and perspectives

The MTBF (Mean Time Between Failures), is at least 4000 hours for an azimuth or elevation system. The production price of such ground equipment is still somewhat uncertain, but will probably be of the order of 2–4 times the price of ILS ground equipment. The costs of ground levelling in front of the antennae, which are considerable in the case of ILS, are, however, far lower for MLS. On the other hand, operational costs of MLS include heating of antenna radoms to remove ice and snow. There is also a probability of a certain amount of restructuring at some airports, principally because ILS and MLS must in some cases coexist during a transition period.

There is widespread scepticism among airline companies with regard to MLS, primarily because of the costs, but also due to uncertainties about technical advantages (Woolley, 1985). An aircraft receiver costs about $25,000, and the airlines also fear that they will be indirectly charged for the costly ground equipment. Because of essential improvements of ILS performance during the last few years (e.g. possibilities of tailoring ground antennae for each airport by means of special algorithms), the relative technical advantages of MLS have been reduced (Breien, 1979). The capacity problems of airports are often caused by runway saturation, and this is not relieved by MLS.

Part II

Satellite systems

7 Satellite orbits and geometry

7.1 Introduction

In order to understand the possibilities and limitations of satellite navigation systems it is necessary to have some knowledge of satellite orbital behaviour. This is also important for understanding the content and processing of navigational messages broadcast from the satellites. A more detailed treatment is given in Appendix 7.

7.2 Kepler's laws

The laws of mechanics governing the movements of the satellites in space stem from the German astronomer Johannes Kepler who studied the orbits of the planets around the sun; but they of course apply also for an artificial satellite moving around the earth. In their simplest form these laws are based on the consideration of masses (the earth and the satellite) as point masses and on the consideration of the active forces as gravitation forces only. It is further assumed that the satellites are not influenced by gravitation forces other than that of the earth, i.e. the distances of other celestial bodies are infinite in comparison. Then Kepler's three laws are valid for the satellite orbits:

1. The satellite orbit is a conical section, i.e. an ellipse, a parabola or a hyperbola, with the gravitation centre of the earth at one of the foci.

2. The vector from the centre of gravity of the earth to the centre of gravity of the satellite sweeps over equal areas in equal times.
3. If the satellite has an elliptic orbit, the square of the orbit time is proportional to the length of the semi-major axis raised to the third power.

The laws derived by Kepler are based on the general gravitational law of Newton:

$$K = G \frac{Mm}{r^2} \tag{7.1}$$

where K is the gravitation force, G is the general gravitation constant ($= 6.67 \times 10^{-11} (m^3/kg)/s^2$), M is the mass of the earth and m that of the satellite, and r is the distance between the centres of gravity of the earth and the satellite. K acts along the line between the centres of gravity. With $M = 5.976 \times 10^{24}$ kg the equation can be abridged: $\mu = GM = 398,600.8$ $(km)^3/s^2$.

7.3 Distortion of the orbits

The deviation between the simple orbit which follows Kepler's laws and the real orbit is, among other things, due to the flattening and irregular mass distribution of the earth, attraction from other celestial bodies, air drag and solar radiation pressure. Among these, the deviation of the earth from the ideal spherical shape is the main reason for an inclined orbit satellite not following the normal orbit.

The undistorted orbit has a constant form, size and spatial orientation, and the centre of gravity would only follow the earth's movement around the sun. However, there are irregularities in this movement as well as in the earth's rotation, and these irregularities are also reflected in the satellite's movement. As an example, the earth's axis does not have a constant direction in space, but is precessing, i.e. describing a conical surface with its peak point at the centre of the earth and an aperture angle of 47° in the course of about 25,800 years. The precession is caused by the fact that the sun does not lie in the equatorial plane of the earth, but in a direction forming an angle of 23.5° to this so that its attraction of the earth is different in the southern and northern hemispheres. The influence from the moon is similar, but as the moon is much smaller, the effect is smaller. The influence of the moon implies that additional motion, nutation, is superimposed on the precession so that the ends of the earth's axis describe serpentine lines in space (Figure 7.1). The nutation has a long periodic variation of 18.6 years due to the rotation of the plane of the moon around the axis of the ecliptic (the ecliptic is the orbital plane of the sun), and a short periodic variation due to the movement of the moon around the earth.

With increasing distance between the earth and the satellite the influence from irregularities in the shape and movement of the earth decreases. For example, at an altitude of about 1000 km the influence of tidal earth on the orbit is about 15 per cent of the direct influence of the sun and the moon. (Tidal earth is an elastic deformation of the earth due to the attraction of the sun and the moon. The magnitude is 20–30 cm.) At the same altitude the air drag and radiation pressure are about equal, but at higher altitudes the air drag and the influence of the tidal earth decrease, whereas the radiation pressure and solar gravitation increase. As an example of the magnitude of the radiation pressure, it may be mentioned that it is about 0.5×10^{-5} N/m^2 on a satellite in a geostationary orbit. A similar effect is given by a reaction force of $F = -P/c$, where P is the transmitted radio power in a certain direction and c is the velocity of light; 1 kW then corresponds to 0.3×10^{-5} N.

As described previously, the theoretical earth surface, the geoid, is defined as the surface formed by the oceans in a state of calm. The surface is imagined to be continued under the continents so that it

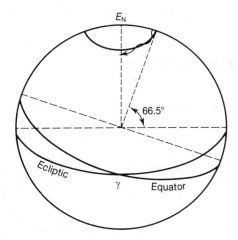

Figure 7.1 Variations in the gravitational forces of the sun and the moon due to the fact that these bodies are not in the equatorial plane imply irregularities in the direction of the earth's axis

forms a closed surface. It is then in balance and always perpendicular to the direction of gravity. This is determined by the mass potential V and by the rotation potential according to

$$W = V + \tfrac{1}{2}\omega^2 r^2 \cos^2 \Phi \qquad (7.2)$$

where ω is the angular velocity of the earth's rotation and Φ is the latitude. A satellite in orbit around the earth is influenced by the potential V so that its orbit can be computed when the variations in V are known. Conversely, observations of satellite orbits and their changes can be utilized for computation of V. The potential can be expanded in a series according to a so-called spherical function, expressing variations over spherical surfaces (whereas, for example, Fourier serial expansions are used for variations in a plane). The first of the coefficients of this serial expansion, J_2, is an expression of the flattening of the earth, and if this is the only coefficient taken into consideration, the potential can be written as

$$V = \frac{\mu}{r} \left[1 + \frac{J_2}{2} \left(\frac{R_E}{2} \right)^2 (1 - 3\sin^2 \Phi) \right] \qquad (7.3)$$

where R_E is the earth's radius in the equatorial plane and r is the distance between the earth's centre and the satellite. The following coefficients give similar terms in addition to Equation (7.3), but J_3, J_4, etc., are at least 1000 times smaller than J_2 and then have a correspondingly smaller influence on the satellite orbit.

7.4 Coordinate systems

To be able to express the position of a satellite in a usable way, a suitable coordinate system is needed. Methods developed by the astronomers to characterize the movements of the planets are also utilized here.

The choice of coordinate system is based on the fact that not only the earth and the sun, but the whole solar system is small compared to the distances to the stars. The distance to the closest star (α Centauri), is about 4 light years ($\approx 4 \times 10^{13}$ km); to *Vega* it is about 30 light years and to the polar star about 50 light years. The sun is 8.25 light minutes (about 150 million km = 1 astronomical unit) from the earth. The stars can therefore be regarded as being located on the inside of a spherical surface of such large dimensions that the whole solar system can be regarded as a point at the centre of the sphere. This sphere is called the *celestial sphere*. Even if the earth can be considered as dimensionless compared to the celestial sphere, the relation between directions on the earth and in the solar system can be transferred to the sphere (Figure 7.2). As a consequence of the large dimensions of the celestial sphere only directions (angles) are considered and not finite distances.

The instantaneous rotational axis of the earth intersects the celestial sphere at the celestial north and south poles (p_n and p_s, respectively, in Figure 7.2). The equatorial plane of the earth can be imagined as expanded to intersect with a celestial sphere. The circle of intersection is then called the *celestial equator*.

The direction of gravity at an observational point on the earth extended upwards intersects the celestial sphere at zenith. When extended downwards it intersects the celestial sphere at nadir. A point of observation on the earth is represented by its zenith point on the celestial sphere.

A great circle passing through the polar points and thus perpendicular to the equator is called a *time circle*.

The time change passing through zenith at a point on the earth's surface is called the *celestial meridian of the point*.

A small circle parallel to the celestial equator is called a *celestial parallel*.

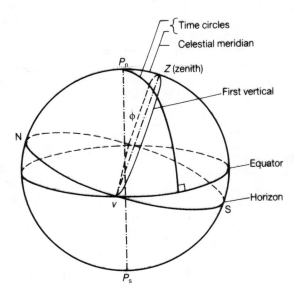

Figure 7.2 The celestial sphere

The horizontal plane at a point of observation extended outwards to the celestial sphere is called the *celestial horizon*.

A plane perpendicular to the horizon passing through zenith is a *vertical plane*. A vertical plane perpendicular to the meridian of the plane of observation is the *first vertical*.

The first vertical intersects the celestial horizon at the east and west points, respectively.

As a consequence of the rotation of the earth around its own axis, the zenith, the vertical plane and the celestial horizon change their locations on the celestial sphere continuously. As the sun rotates from the west towards the east, this gives the impression that the celestial sphere rotates from the east towards the west, and that the stars describe circles parallel to the equator.

The earth moves in an elliptic orbit (the ecliptic) around the sun (Figure 7.3). (The sun appears to orbit the earth.) The ecliptic forms an angle of about 23°27′ to the celestial equator. The intersection points between those two are the vernal and autumnal equinoxes respectively. The centre of gravity of the sun (the solar centre) has been chosen as the centre of the celestial sphere. Due to the large dimension of the celestial sphere, a direction, for example from the solar centre to the celestial north pole, is parallel to the direction from the centre of the earth to the celestial north pole.

A coordinate system in common use to denote the movement and position of satellites is called the *rectascension system*, and its elements are called the *elements of Kepler* (Figure 7.4). There are six such elements, and together they completely characterize the position of the satellite. The need for six elements instead of three results

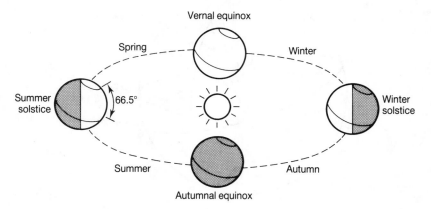

Figure 7.3 The orbit of the earth around the sun

from the fact that the satellite movement is absolute as well as relative. The satellite orbit moves along with the earth in the solar orbit of the earth, and the satellite moves on the orbit. It is therefore necessary to have three elements to denote the position of the satellite orbit in space, and three elements to describe the shape of the orbit and the position of the satellite itself in the orbit. It should be emphasized that this is not the only way of characterizing a satellite's position. One might well use, for example, a Cartesian earth-fixed system.

The advantage of using Kepler elements is that the satellite position is easy to extrapolate once the receiver knows these elements and (if necessary) their time derivatives. On the other hand, Cartesian coordinates are more easily used for receiver position calculation.

In Figure 7.4 the direction to the vernal equinox is the reference direction. The origin of the system is the centre of gravity of the sun, but because of the size of the celestial sphere, the directions from a point on the earth, or in the earth orbit, are the same. The angle Ω in the equatorial plane measured eastwards from the direction to the vernal equinox and to the direction of the point, as seen from the

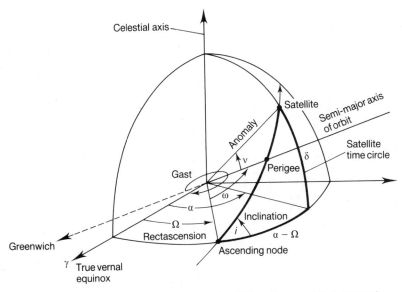

Figure 7.4 Keplerian orbital parameters (GAST = Greenwich Apparent Sidereal Time): Ω = rectascension of ascending node; i = inclination; ω = argument of perigee; a = semi-major axis; e = eccentricity of the satellite orbit; ν = true anomaly of the satellite

centre of the earth, where the satellite orbit passes through the equatorial plane on its way from the south to the north, is called the *rectascension of the ascending node*. Note that this is a fixed angle which is independent of the earth's rotation because the satellite orbit is fixed in relation to the stars, while the earth rotates within the satellite orbit. (As mentioned above, certain external forces can influence the satellite orbit which, in that case, changes slightly in space.) A slope angle, i, between the orbital plane of the satellite and the equatorial plane is called the *inclination*. The angles Ω and i together consequently describe the orientation of the orbital plane of the satellite in space.

As the satellite orbit is an ellipse, another parameter is needed to describe the orientation of the orbit in the orbital plane. This angle, ω in Figures 7.4 and 7.5, is called *argument of the perigee*. This is the angle in the orbital plane between the direction of the major axis and the direction to the ascending node.

The satellite orbit itself is shown by Figure 7.5. The perigee is that point on the orbit where the satellite is closest to the earth, the opposite point is called the *apogee*.

In order to characterize the position of the satellite in the orbit at a

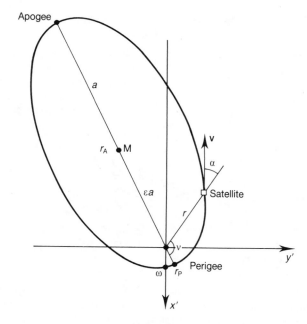

Figure 7.5 Satellite orbit

certain point of time, the angle v between the semi-major axis in the direction of perigee and the vector to the satellite is used. This angle is called the *true anomaly*. (The word 'true' marks that this is the value at a certain point of time and not an average value.) The orbital parameters of the satellite must also be tied to a terrestrial coordinate system for use in terrestrial navigation. Transit and GPS here use WGS 84.

8 Satellite navigation principles

8.1 Frequency measurements

It is a well-known fact that the relative movements between a transmitter and a receiver introduce frequency changes or Doppler shifts in a received signal. The phenomenon is utilized for self-contained aircraft and helicopter navigation where radar on board transmits signals in well-defined directions (referenced to the aircraft), measuring the frequency change of the signals reflected back from the ground below. As the Doppler shift is directly proportional to the radial velocity between the transmitter and the receiver, the position change in a time interval can be measured by integrating the Doppler shift over this interval.

For satellite navigation systems this principle opens a possibility that is usually non-existent for terrestrial systems, i.e. the utilization by the receiver of the Doppler shift caused by the movement of the transmitter. As the relative radial velocity, v_r, gives rise to the Doppler shift

$$f_d = \frac{v_r}{\lambda} \tag{8.1}$$

a possible movement of the navigational receiver is only a complication. The method is thus most suitable for stationary receivers.

The Doppler shift of a signal received on the ground from a satellite orbiting the earth behaves as shown in Figure 8.1. A receiver on the satellite orbit itself records a Doppler shift according to curve No. 1, i.e. f_d is positive (and constant for the closest part of the orbit

204

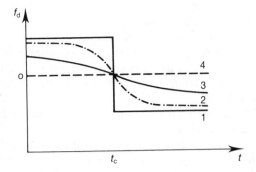

Figure 8.1 Doppler shift in satellite signals

which approximates to a straight line) while the satellite is approaching and changes its sign once past the receiver.

The more distant the receiver is from the point of closest approach of the satellite orbit, the smaller is the slope of the curve. The curves of Figure 8.1 are thus numbered with increasing distance to this point, and curve No. 4 denotes infinite distance. The shape of the curve is consequently directly dependent on this distance, and when the satellite orbit is known, the position of the receiver can be determined from the shape of the curve. In practice (Chapter 10), this is done by assuming a receiver position, computing a resulting curve and comparing the results with actual measurements. After that the assumed position is changed until the rms deviation between the computed and measured curve shapes is less than a threshold value (method of least squares, Appendix 5).

On each side of a satellite orbit, as seen from the earth, there is such a minimum point. Because of the earth's rotation, however, the Doppler curve is measurably different at the two points. This fact can be utilized if the knowledge of the user's own position is not good enough to prevent the ambiguity.

As the result from measurements of the Doppler shift of a received satellite signal depends only on the carrier frequency of the satellite and the orbital velocity, as well as on the angle between the orbit and the direction to the user from the satellite at the instant of measurement, the surface of position is a cone with its top at the satellite and half the top angle equal to the above angle. The intersection between such cones and the earth's surface is, if the earth is regarded as a plane over a small area, a hyperbola. Another line of position is needed for complete position determination in the plane, and a new

Doppler measurement gives this, either simultaneously from measurements of signals from another satellite or measurements of the former satellite's signals at another point in time.

In practice, a relatively large number of measurements is made for each satellite so that the starting point of the position computations is an overdetermined system of equations to be solved by matrix methods (Appendix 5).

Another method is to measure the phase variation instead of the frequency variation. This implies that the Doppler shift of the received signals is integrated over a known time interval; the result is proportional to the distance change between the satellite and the receiver during this interval. The surface of position in this case is a hyperboloid. (The method is utilized in TRANSIT, Chapter 10, p. 225ff.)

8.2 Time measurements

Measurements of time differences between a transmitted and received signal which are then translated into a distance between the satellite transmitter and a receiver is, in principle, a very simple and straightforward method. The reason for not using this method in the first navigational satellites was the need for a good time reference to give good accuracies.

The surfaces of position for distance measurements are spherical surfaces, and their intersections of a plane or other spherical surfaces give circles (Figure 8.2). If bias errors are not considered, e.g. the receiver's own clock errors, there is a considerable probability that the error volume (stippled in Figure 8.2) does not include the correct point.

8.3 Measurement principles

As mentioned in Section 8.1, systems using frequency measurements either measure the Doppler shift itself, or its integral. Doppler frequency measurements are often based on FFT (Fast Fourier Transforms) (Forssell, 1980). More common, however, is the method of integrated Doppler which is described in more detail in Chapter 10. This latter method is also used for aiding direct distance measure-

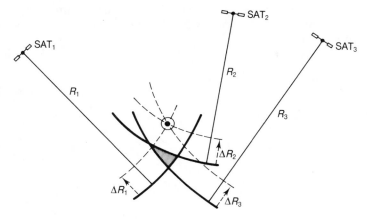

Figure 8.2 Distance measurement errors give an error volume with spherical boundary surfaces

ments, e.g. in GPS (Chapter 12). Integrating the Doppler shift means measuring the phase change during the integration time interval. This change can also be measured directly by tracking the phase of the received carrier. Loss of track may be a problem when using this technique (Chapter 12). As described in Section 8.1, frequency measurements give conical surfaces of position, whereas phase change measurements give hyperboloids. Intersection of a plane earth surface gives hyperbolae in both cases.

Direct distance measurements are carried out as time-of-arrival (TOA) measurements or phase measurements as in terrestrial systems. TOA is usually measured by means of correlating or matched-filter receivers using receiver-generated replicas of known satellite signal modulation forms (codes). (Details are given in Chapter 11.) Also, distance differences are measured, simply by subtracting distance measurements at different times. The difference obtained should be equal to the corresponding carrier-phase change in the measurement time interval. As the carrier phase can be tracked with a considerably smaller range error, this equality can be used for smoothing direct distance measurements, thus improving accuracy.

For high-accuracy (submetre) relative positioning, phase differences are measured with regard to a reference point. Because of various problems (ambiguities, phase slips, etc., see Section 12.8), post-processing of measurement results is usually necessary.

Finally, it should be mentioned that receiver velocity can also be measured by direct distance systems, either by Doppler or distance difference measurements. As satellite-to-user time differences must be determined, these systems can also be used for time transfer between known positions.

⑨ Error calculations

9.1 Error sources

Physically, the external error sources of time, as well as frequency measurements, are largely the same and can be treated in a similar way. The different error sources may, however, have different influences on the accuracies of positioning or navigation in the two cases.

A most important error source is the uncertainty with regard to the satellite's own position, i.e. variations in orbital parameters. All navigation satellites transmit information about their positions to the user, but this information is always contaminated with certain errors. These have three main causes:

(a) difficulties in determining the position of the satellite from the ground;

(b) irregularities in the orbital parameters of the satellite so that the result of a determination of the orbit is valid for a limited amount of time only;

(c) limited capacity of information transfer from the satellite to the user.

It has been explained in the preceding chapter why the satellite orbits are not as regular and predictable as might be concluded by the laws of Kepler. The accuracy of an orbit determination is also necessarily limited, but so-called post-processing, i.e. coordination and processing of measurement results from many earth stations tracking the same satellite usually gives results with accuracies of the order of a metre. These data, however, are only available afterwards, so a navigating user has to be content with the lower accuracy given directly from the satellite.

The most essential error source, other than the orbital parameters, is usually the propagation conditions. In this respect, however, satellite systems have advantages compared to most terrestrial systems because there is only one signal path, uninfluenced by the varying electric properties of the earth's surface. (The exception is reflections from the neighbourhood of the receiving antenna.) The influence of the ionosphere is not negligible but, because of the frequency choice, it can be managed. It was explained in Chapter 3 how the ionosphere can be considered as a dielectric with a refraction index:

$$n = \sqrt{1 - (f_p/f)^2} \tag{9.1}$$

where f_p is the plasma frequency, usually in the range 2–20 MHz. As the carrier frequencies of the satellites are at least ten times higher than f_p, n can be serially expanded, and if only the first term is considered, we have

$$n \approx 1 - \frac{1}{2}\left(\frac{f_p}{f}\right)^2 \tag{9.2}$$

The difference between the refraction index (i.e. the phase velocity), in a vacuum and in the ionosphere is consequently inversely proportional to the square of the frequency. For that reason the navigational satellites usually transmit on two frequencies, which makes it possible to determine the constant of proportionality and, consequently, remove the influence of the ionosphere to a large extent. For different reasons, however, many satellite receivers can receive one frequency only, leaving the ionospheric error as it is if mathematical models to reduce the influence are not utilized (Section 12.5.3.3).

The ionospheric delay (Equations (3.37) and (9.2)) can be written as

$$\Delta t = \int_l \frac{dl}{c}\,(n - 1) \approx \int \frac{1}{2c}\,(f_p/f)^2\,dl = \frac{k}{f^2} \tag{9.3}$$

(where l is the signal path). For systems using two carrier frequencies, f_1 and f_2, we have two different delays, Δt_1 and Δt_2, and a two-frequency receiver measures the difference in signal time-of-arrival (TOA), $\Delta t_{2,1}$, where

$$\Delta t_{2,1} = k(f_2^{-2} - f_1^{-2}) \tag{9.4}$$

Thus

$$k = \frac{\Delta t_{2,1} f_1^2 f_2^2}{f_1^2 - f_2^2} \tag{9.5}$$

and

$$\Delta t_1 = \frac{\Delta t_{2,1}}{(f_1/f_2)^2 - 1} \tag{9.6}$$

Using Equations (3.36) and (9.3) we obtain

$$\Delta t_1 \approx \frac{q_e^2}{8\pi^2 c f_1^2 \varepsilon_0 m_e} \int_l N_e \, dl \tag{9.7}$$

where

$$\int_l N_e \, dl = TEC \tag{9.8}$$

i.e. the total free electron content in a column with 1 m^2 cross-section along the whole signal path.

As has been indicated before, the TEC value varies because of the properties of the ionosphere, but also as a function of the obliqueness of the signal path. Because of the unsurmountable task of determining the variation of N_e along the path, it is often assumed (Sigmond, 1990) that the ionosphere can be regarded as a homogeneous shell having a certain height, e.g. 350 km, as in Klobuchar (1987). For computations that are in accordance with Equations (9.6)–(9.8), this is of no significance, but it does play a role in estimating the residual error of the two-frequency method.

There are many investigations of TEC as a function of the position and time of year and day, but there is still a lack of knowledge concerning ionospheric behaviour, particularly at high latitudes. In general, it can be said that TEC is usually smaller at high than at low and middle latitudes, but the fluctuation is more rapid as a function of time and position.

If N_e is assumed constant from the receiver to an altitude of h_0, the signal path length through the ionosphere is (Figure 9.1)

$$l = R\left[\sqrt{\sin^2 E + \left(\frac{h_0}{R}\right)^2 + 2\,\frac{h_0}{R}} - \sin E\right] \tag{9.9}$$

where R is the radius of the earth. A constant N_e along this distance gives (Equations (9.3) and (9.7))

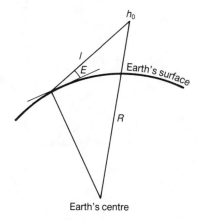

Figure 9.1 Calculation of the oblique distance to the top of a homogeneous ionosphere

$$\Delta t_1 = -\frac{q_e^2 N_e l}{8\pi^2 c m_e \varepsilon_0 f_1^2} \tag{9.10}$$

i.e.

$$k = -\frac{q_e^2 N_e l}{8\pi^2 c m_e \varepsilon_0} \tag{9.11}$$

Assuming that $f_p \gg f$ as in Equation (9.3), such that the second term in the series expansion, which is

$$\tfrac{1}{8}(f_p/f_1)^4 \ll \tfrac{1}{2}(f_p/f_1)^2$$

we get

$$\Delta t_1 = \int_0^l \left[\sqrt{1 - \frac{N_e q_e^2}{4\pi^2 f_1^2 m_e \varepsilon_0}} - 1 \right] \frac{dl}{c}$$

$$\tag{9.12}$$

$$\approx \frac{l}{c} \left[\sqrt{1 + \frac{2kc}{f_1^2 l}} - 1 \right]$$

The residual error of the two-frequency method for TOA measurements is then

$$\Delta t_{1\,\text{err}} = \frac{l}{c} \left[\sqrt{1 + \left(\frac{2kc}{f_1^2 l}\right)} - 1 \right] - \frac{k}{f_1^2} \tag{9.13}$$

where k is given by Equation (9.5) and l by Equation (9.9).

It should be emphasized that Equation (3.38), and consequently Equation (9.3), is valid only for phase velocity (v_f) calculations, i.e. calculations pertaining to carrier phase. The group velocity is

$$v_g = \left[\frac{d\beta}{d\omega}\right]^{-1} \tag{9.14}$$

where

$$\beta = 2\pi/\lambda = \omega/v_f = \omega n/c \tag{9.15}$$

Equations (9.1), (9.14) and (9.15) give

$$v_g = \left[\frac{d}{d\omega}(\sqrt{\omega^2 - \omega_p^2})/c\right]^{-1} = cn \tag{9.16}$$

Thus,

$$v_g v_f = c^2 \tag{9.17}$$

and TOA errors caused by the ionosphere have different signs depending on whether carrier or modulation is measured.

As mentioned at the end of Chapter 3, the influence of the collision frequency and of the earth's magnetic field has been neglected (Equations (3.37) and (9.1)). However, when calculating the residual error after the two-frequency correction, the gyrofrequency term could be included. This modifies Equation (9.13) slightly as the term under the root sign should be increased by $\omega_p^3 \omega_H \cos \theta/\omega^2$ (from a serial expansion of Equation (3.39)). However, it should still be kept in mind that the solution is an approximation.

The troposphere has a similar influence on satellite systems as it has on terrestrial systems, with one exception: most of the signal path is above it. The influence can be calculated from formulas presented in Section 3.4. In addition, satellites at least 5–10° above the horizon are preferred, which reduces the influence of both the ionosphere and the troposphere compared to the case of smaller elevation angles.

The troposphere also causes a delay of the signals, but this delay is largely independent of frequency (to about 30 GHz). The time variability of this delay makes it impractical to determine the influence accurately, so in general, approximate empirical formulas are used with average parameter values (Section 3.4). For altitudes up to about 1 km, the refractivity decreases linearly with the altitude (Equation (3.28)), whereas for higher altitudes the decrease is exponential (Equation (3.27)). For altitudes above 8–9 km the total delay caused by the troposphere for vertical incidence is usually less than 2.5 ns (NATO, 1987). The average delay for vertical incidence can thus be calculated as

$$\Delta t = \Delta t_1 + \Delta t_2 + \Delta t_3 \qquad (9.18)$$

where

$$\Delta t_1 = \frac{1}{c} \int_0^{1000} (N_s - h\Delta N)\, dh \qquad (9.19)$$

$$\Delta t_2 = \frac{1}{c} \int_{1000}^{h_{ref}} \left(N_s - \frac{10^6}{2} \Delta N \right) e^{-qh}\, dh \qquad (9.20)$$

$$\Delta t_3 = 2.5 \text{ (ns)}$$

(See Section 3.4 for definitions.) For oblique signal incidence, the above value of Δt should be multiplied by an obliquity factor containing the elevation angle to the satellite,

$$f_{obl}(E) = \frac{h_{ref}}{R\left[\sqrt{\sin^2 E + \left(\frac{h_{ref}}{R}\right)^2 + \frac{2h_{ref}}{R}} - \sin E \right]} \qquad (9.21)$$

in analogy with Equation (9.9). ($h_{ref} \approx 9000$ m).

Satellite systems also have a number of internal error sources of more or less random character. As mentioned, the clock error (of the receivers) is essential, but this can be taken care of as an extra unknown in the navigational computations. These various error sources and their influences are described in connection with the respective system. This also applies to the receiver velocity error which may be dominant in frequency systems if not compensated for, as receiver movement causes Doppler shifts in analogy with satellite movement.

9.2 The geometrical influence on the accuracy

9.2.1 Frequency measurements

It is assumed that both the satellite and receiver are referred to the same (terrestrial) Cartesian coordinate system and have the coordinates (x_i, y_i, z_i) and (x_u, y_u, z_u) respectively. The distance between them is then

$$D_i = [(x_u - x_i)^2 + (y_u - y_i)^2 + (z_u - z_i)^2]^{1/2} \qquad (9.22)$$

The Doppler shift at the receiver is (the receiver does not move)

$$f_{di} = \frac{v_{ri}}{\lambda} = -\frac{\dot{D}_i}{\lambda} = \frac{\dot{x}_i(x_u - x_i) + \dot{y}_i(y_u - y_i) + \dot{z}_i(z_u - z_i)}{D_i\lambda}$$

$$(9.23)$$

where λ is the wavelength of the satellite signal. (In $\dot{D}_i$ the movement of the satellite in the orbit and the earth's rotation are included.) When Doppler shift measurements have been carried out, the desired position is given by a system consisting of, at least, three equations of the same type as Equation (9.23). With vector notation, this can be written

$$f_{di} = \dot{\mathbf{D}}_i \cdot (\mathbf{R}_u - \mathbf{R}_i)/D_i\lambda \qquad (9.24)$$

where $\mathbf{R}_i$ is the ith satellite position, $\mathbf{R}_u$ is the user position and f_{di} denotes the ith Doppler measurement $(i \geq 3)$. The solution to this system gives the user position (x_u, y_u, z_u) as an implicit function, F, of the measurement results f_{di}, the satellite positions (x_i, y_i, z_i), and the wavelength, λ, i.e.

$$F(x_u, y_u, z_u, f_{di}, x_i, y_i, z_i, \lambda) = 0 \qquad (9.25)$$

The errors of the position determination are consequently obtained by differentiating the function with regard to these quantities (Forssell, 1980):

$$\frac{\partial F}{\partial x_u} \Delta x_u + \frac{\partial F}{\partial y_u} \Delta y_u + \frac{\partial F}{\partial z_u} \Delta z_u + \frac{\partial F}{\partial f_{di}} \Delta f_{di} + \frac{\partial F}{\partial x_i} \Delta x_i$$

$$+ \frac{\partial F}{\partial y_i} \Delta y_i + \frac{\partial F}{\partial z_i} \Delta z_i + \frac{\partial F}{\partial \lambda} \Delta \lambda = 0 \qquad (9.26)$$

Satellite position uncertainty is often given as along-track, cross-track and radial errors, but there is usually an unknown correlation between these errors, so it may be appropriate to treat the satellite position error as the distance, ΔD_i, between the stated and the real radial positions. Then, Equation (9.26) can be simplified somewhat:

$$\frac{\partial F}{\partial x_u} \Delta x_u + \frac{\partial F}{\partial y_u} \Delta y_u + \frac{\partial F}{\partial z_u}$$

$$+ \frac{\partial F}{\partial f_{di}} \Delta f_{di} + \frac{\partial F}{\partial D_i} \Delta D_i + \frac{\partial F}{\partial \lambda} \Delta \lambda = 0 \qquad (9.27)$$

where $D_i = |\mathbf{R}_i - \mathbf{R}_u|$, and the quantities Δf_{di}, ΔD_i and $\Delta \lambda$ are uncorrelated.

The rms errors of the frequency measurements, of the positions of the satellites and of the wavelengths are assumed to be σ_{fd}, σ_s and σ_λ respectively.

In order to calculate error ellipses in the horizontal plane, which is often desirable, Equation (9.27) is squared and averaged, and the result is equations of the form (Forssell, 1980)

$$a_i^2 \sigma_x^2 + b_i^2 \sigma_y^2 + 2a_i b_i \sigma_x \sigma_y \rho_{xy} = k_{fi}^2 \sigma_{fd}^2 + k_{si}^2 \sigma_s^2 + k_\lambda^2 \sigma_\lambda^2 \qquad (9.28)$$

a_i, b_i, k_{fi}, k_{si}, and k_λ are constants containing the derivatives with regard to x_u, y_u, f_d, D_i and λ, respectively, and $\rho_{xy} = \overline{\Delta x_u \Delta y_u}/\sigma_x \sigma_y$. The influence by the geometry is contained in the coefficients a_i, b_i, etc. The system of Equation (9.28) is solved according to the method of least squares with regard to σ_x^2, σ_y^2, and ρ_{xy}. After that the joint density function of the position errors in the x and y directions is formed (Section 2.4):

$$p(\Delta x_u, \Delta y_u)$$

$$= \frac{\exp\left\{-\frac{1}{2(1 - \rho_{xy}^2)} \left[\left(\frac{\Delta x_u}{\sigma_x}\right)^2 + \left(\frac{\Delta y_u}{\sigma_y}\right)^2 - 2\rho_{xy} \frac{\Delta x_u \Delta y_u}{\sigma_x \sigma_y}\right]\right\}}{2\pi\sigma_x \sigma_y \sqrt{1 - \rho_{xy}^2}}$$

$$(9.29)$$

The ellipse described by Equation (9.29) when the exponent is constant can now be calculated with regard to its size and orientation.

It is usually most suitable to utilize low-orbit satellites (i.e. satellites orbiting in the order of 1000 km above the ground) for Doppler positioning as these have the greatest velocity related to the users (Equation (A7.40)). Examples of such satellite systems are TRANSIT and the search-and-rescue system COSPAS/SARSAT (Stansell, 1978; Scales and Swanson, 1984).

Equations (9.25)–(9.29) are difficult to visualize geometrically, but have been used for comprehensive computations of error ellipses with realistic values of σ_{fd}, σ_s and σ_λ (Forssell, 1980). It appears that the frequency measurement is often the dominating source of error. When this is the case, the error ellipses are elongated with a cross-orbit major axis around the point $f_d = 0$. (It is advantageous to include this point as $|\dot{f}_d|$ has its maximum value there, so that is where the different satellite orbits differ most.) For satellites passing through zenith, the error ellipses have eccentricities close to unity, i.e. almost infinitely large cross-track errors, which is obvious as a small displacement of the receiver in that direction gives a minute change of the Doppler curve.

If the frequency error does not dominate, the major axis of the error ellipse is usually not perpendicular to the satellite orbit.

For small elevations, orbit errors and ionospheric/tropospheric errors give elongated ellipses in the orbit direction.

9.2.2 Time measurements

9.2.2.1 Calculation of geometrical dilution of position

For measurements of times of arrival of satellite signals, i.e. determination of distances between a satellite and the user, the geometrical position of the user in relation to the satellite plays a considerable role, as it does for Doppler measurements. As for terrestrial systems, the final position error is a function of the distance measurement error and a geometrically determined factor. When more satellites than necessary are available at the same time, i.e. are found over the least acceptable elevation angle (usually 5° for high-orbit satellites), the resulting accuracy is influenced by choosing the most favourably positioned satellites. For this reason the following presentation is relatively detailed so that the different aspects are thoroughly illustrated.

From Figure 9.2 we have (Milliken and Zoller, 1978)

$$\mathbf{R}_u = \mathbf{R}_i - \mathbf{D}_i \tag{9.30}$$

$\mathbf{R}_u$ is the position vector of the user and $\mathbf{R}_i$ that of the ith satellite, with origin at the earth's centre, and $\mathbf{D}_i$ is the vector from the receiver to the ith satellite, $1 \le i \le n$, $n \ge 4$. The components of $\mathbf{R}_i$ are known from the orbital data of the satellite. An interesting quantity is the length of $\mathbf{D}_i$, and by defining the unity vector $\mathbf{e}_i$ along

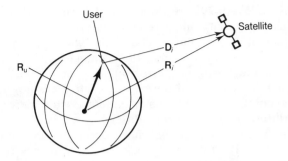

Figure 9.2 Geometry for the calculation of the position of the user

$\mathbf{D}_i$ this length can be calculated according to Equation (9.30):

$$D_i = \mathbf{e}_i \cdot \mathbf{D}_i = \mathbf{e}_i \cdot \mathbf{R}_i - \mathbf{e}_i \cdot \mathbf{R}_u \tag{9.31}$$

It can also be written,

$$D_i = \rho_i - B_u + B_i \tag{9.32}$$

where ρ_i is the distance measured by the receiver, and B_u and B_i are the clock errors of the receiver and of the satellite respectively (expressed as distances). The combination of Equations (9.31) and (9.32) gives

$$\mathbf{e}_i \cdot \mathbf{R}_u - B_u = \mathbf{e}_i \cdot \mathbf{R}_i - \rho_i - B_i \tag{9.33}$$

which with $i = 1, 2, \ldots, n$ is the set of equations providing solutions to the four unknowns x_u, y_u, z_u and B_u. These can be utilized to define the matrix

$$\mathbf{X}_{u(4\times1)} = (x_u, y_u, z_u, B_u)^{\mathrm{T}} \tag{9.34}$$

(The parenthesis of the matrix index contains the number of lines and columns, respectively, of the matrix.)

By projecting the mentioned unity vectors $\mathbf{e}_i$ on to the three coordinate axes we define the matrix

$$\mathbf{\Gamma}_{i(1\times4)} = (e_{ix}, e_{iy}, e_{iz}, -1) \tag{9.35}$$

and the matrices

$$\mathbf{G}_{u(n\times4)} = \begin{bmatrix} \mathbf{\Gamma}_1 \\ \mathbf{\Gamma}_2 \\ \vdots \\ \mathbf{\Gamma}_n \end{bmatrix} \tag{9.36}$$

and

$$\mathbf{A}_{u(n\times4n)} = \begin{bmatrix} \mathbf{\Gamma}_1 & 0 & 0 & \cdots & & 0 \\ 0 & \mathbf{\Gamma}_2 & 0 & & & 0 \\ 0 & 0 & \mathbf{\Gamma}_3 & & & 0 \\ \vdots & & & & & \\ 0 & & & & & \mathbf{\Gamma}_n \end{bmatrix} \tag{9.37}$$

where $\mathbf{0}$ is the null matrix $(0, 0, 0, 0)$. By projecting the position vectors of the satellites on to the coordinate axes we get the components x_i, y_i and z_i, and the matrix

$$\mathbf{S}_{(4n\times1)}(x_1, y_1, z_1, B_1, x_2, y_2, z_2, B_2, \ldots)^{\mathrm{T}} \tag{9.38}$$

In the same way a matrix with the measurement values ρ_i is defined

$$\boldsymbol{\rho}_{(n\times1)} = (\rho_1, \rho_2, \ldots, \rho_n)^{\mathrm{T}} \tag{9.39}$$

The set of Equations (9.33) is then written (with $i = 1, 2, \ldots, n$)

$$\mathbf{G}_u\mathbf{X}_u = \mathbf{A}_u\mathbf{S} - \boldsymbol{\rho} \tag{9.40}$$

If $n > 4$, the system (9.40) is overdetermined, and the solution is found by using the method of least squares (Appendix 5). Thus the solution is

$$\mathbf{X}_u = (\mathbf{G}_u^{\mathrm{T}}\mathbf{G}_u)^{-1}\mathbf{G}_u^{\mathrm{T}}(\mathbf{A}_u\mathbf{S} - \boldsymbol{\rho}) \tag{9.41}$$

Note that the matrices $\mathbf{G}_u$ and $\mathbf{A}_u$ mainly consist of the coefficients e_{ij} which are the cosines of the angles between the directions towards the satellites and the coordinate axes, the so-called directional cosines. As Equation (9.41) is non-linear, because the directional cosines are functions of the desired position, the system requires an iterative solution. Therefore an initial estimate of our own position has to be entered to enable us to calculate these directional cosines, which then give a new position, etc. Because of the large distance to the satellites, the directions to these vary quite slowly, however, when changing the position on the earth (200 km on the earth gives around 0.5° of directional change, for example). Thus only a few iterations are needed to achieve a desired accuracy of computation.

When computing position errors as functions of distance measurement errors, $\mathbf{X}_u$ is differentiated with regard to $\boldsymbol{\rho}$, and in this case it is suitable to regard the matrices $\mathbf{G}_u$ and $\mathbf{A}_u$ as constants:

$$\delta\mathbf{X}_u = (\mathbf{G}_u^T\mathbf{G}_u)^{-1}\mathbf{G}_u^T\delta(\mathbf{A}_u\mathbf{S} - \boldsymbol{\rho}) \qquad (9.42)$$

The mean square error matrix is then the covariance of $\delta\mathbf{X}_u$, and thus

$$\begin{aligned}
\mathrm{cov}\,(\delta\mathbf{X}_u) &= E[\delta\mathbf{X}_u \cdot \delta\mathbf{X}_u^T] \\
&= (\mathbf{G}_u^T\mathbf{G}_u)^{-1}\mathbf{G}_u^T\mathrm{cov}\,[\delta(\mathbf{A}_u\mathbf{S} - \boldsymbol{\rho})][(\mathbf{G}_u^T\mathbf{G}_u)^{-1}\mathbf{G}_u^T]^T
\end{aligned}$$
$$(9.43)$$

The covariance matrix of $\delta(\mathbf{A}_u\mathbf{S}_u - \boldsymbol{\rho})$ is the statistical error of the distance measurements (satellite orbit errors, propagation errors and instrument errors), whereas the other factors of Equation (9.43) only reflect the geometry of the system. If the distance errors of the satellite measurements are mutually independent and have equal variances σ_0^2, we obtain

$$\mathrm{cov}\,[\delta(\mathbf{A}_u\mathbf{S} - \boldsymbol{\rho})] = \sigma_0^2\mathbf{I} \qquad (9.44)$$

where $\mathbf{I}$ is the identity matrix. Then Equation (9.43) after simplification is ($\mathbf{G}_u^T\mathbf{G}_u$ is symmetrical)

$$\mathrm{cov}\,(\delta\mathbf{X}_u) = (\mathbf{G}_u^T\mathbf{G}_u)^{-1}\,\sigma_0^2 \qquad (9.45)$$

Note that, after all, Equation (9.45) is only an approximation (albeit a very good approximation), as the geometry is also part of the matrix $\mathbf{A}_u$. The explicit form of Equation (9.45) is

$$(\mathbf{G}_u^T\mathbf{G}_u)^{-1}\sigma_0^2 = \begin{bmatrix} \sigma_{xx}^2 & \sigma_{xy}^2 & \sigma_{xz}^2 & \sigma_{xt}^2 \\ \sigma_{yx}^2 & \sigma_{yy}^2 & \sigma_{yz}^2 & \sigma_{yt}^2 \\ \sigma_{zx}^2 & \sigma_{zy}^2 & \sigma_{zz}^2 & \sigma_{zt}^2 \\ \sigma_{tx}^2 & \sigma_{ty}^2 & \sigma_{tz}^2 & \sigma_{tt}^2 \end{bmatrix} \sigma_0^2 \qquad (9.46)$$

where the diagonal values are the variances of the errors of the estimated position along the three coordinate axes, and of the estimated time. The root sum square of these four variances is called GDOP (Geometrical Dilution of Precision), and gives the multiplication factor of the estimated distance measurement error for the estimation of total position and time errors (Equation (9.47)). The corresponding error factors can be defined from Equation (9.46) for the total position, horizontal, vertical and time errors, respectively (Equations (9.48)–(9.51)).

$$GDOP = \sqrt{\sigma_{xx}^2 + \sigma_{yy}^2 + \sigma_{zz}^2 + \sigma_{tt}^2} \tag{9.47}$$

$$PDOP = \sqrt{\sigma_{xx}^2 + \sigma_{yy}^2 + \sigma_{zz}^2} \tag{9.48}$$

$$HDOP = \sqrt{\sigma_{xx}^2 + \sigma_{yy}^2} \tag{9.49}$$

$$VDOP = \sigma_{zz} \tag{9.50}$$

$$TDOP = \sigma_{tt} \tag{9.51}$$

σ_{ij}, where $i \neq j$, is the correlation between the errors in the ith and jth directions.

As GDOP is the root sum square of the variances along the coordinate axes, the quantity can be regarded as the distance between two (fixed) points in space and, consequently, it is independent of the selection of coordinate system. (TDOP is of course independent of this selection, too.)

The variation of GDOP as a function of the satellite constellation is examined in the following (for simplification, the number of satellites is limited to four) (Kihara and Okada, 1984).

In Figure 9.3 unity vectors e_1–e_4 point from the position of the user towards the satellites with corresponding numbers forming a pyramid or a tetrahedron with a base area between the points of the vectors e_1–e_3 and the top at the point of e_4. The base area is $\frac{1}{2}|A \times B|$ and the volume of the whole tetrahedron is

$$V = \tfrac{1}{6}(A \times B) \cdot C \tag{9.52}$$

As

$$A = e_2 - e_1 = (e_{2x} - e_{1x}, e_{2y} - e_{1y}, e_{2z} - e_{1z}) \tag{9.53}$$

$$B = e_3 - e_2 = (e_{3x} - e_{2x}, e_{3y} - e_{2y}, e_{3z} - e_{2z}) \tag{9.54}$$

$$C = e_4 - e_3 = (e_{4x} - e_{3x}, e_{4y} - e_{3y}, e_{4z} - e_{3z}) \tag{9.55}$$

Equation (9.52) can be written as a determinant:

$$V = \frac{1}{6} \begin{vmatrix} e_{2x} - e_{1x} & e_{2y} - e_{1y} & e_{2z} - e_{1z} \\ e_{3x} - e_{2x} & e_{3y} - e_{2y} & e_{3z} - e_{2z} \\ e_{4x} - e_{3x} & e_{4y} - e_{3y} & e_{4z} - e_{3z} \end{vmatrix} \tag{9.56}$$

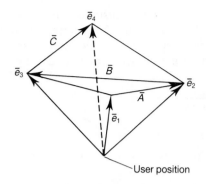

Figure 9.3 Definition of the tetrahedron

The determinant of the matrix $\mathbf{G}_u$ (Equation (9.36)) is

$$
|\mathbf{G}_u| = \begin{vmatrix}
e_{1x} & e_{1y} & e_{1z} & -1 \\
e_{2x} & e_{2y} & e_{2z} & -1 \\
e_{3x} & e_{3y} & e_{3z} & -1 \\
e_{4x} & e_{4y} & e_{4z} & -1
\end{vmatrix}
\tag{9.57}
$$

For determinants it is known that a row can be multiplied by an arbitrary constant and added to another row without changing the value of the determinant. If the third row is subtracted from the fourth, the second row from the third and the first row from the second in this order, and the determinant is solved along the fourth column (which then has the form $(-1\ 0\ 0\ 0)^T$), the result is

$$
|\mathbf{G}_u| = \begin{vmatrix}
e_{2x} - e_{1x} & e_{2y} - e_{1y} & e_{2z} - e_{1z} \\
e_{3x} - e_{2x} & e_{3y} - e_{2y} & e_{3z} - e_{2z} \\
e_{4x} - e_{3x} & e_{4y} - e_{3y} & e_{4z} - e_{3z}
\end{vmatrix}
\tag{9.58}
$$

The determinant $|\mathbf{G}_u|$ is thus proportional to the volume V (Equation (9.56)).

Without loss of generality the x-axis can be positioned along $\mathbf{e}_1$, and $\mathbf{e}_2$ is positioned in the xy plane (Figure 9.4). Then we have

$$
|\mathbf{G}_u| = \begin{vmatrix}
1 & 0 & 0 & -1 \\
e_{2x} & e_{2y} & 0 & -1 \\
e_{3x} & e_{3y} & e_{3z} & -1 \\
e_{4x} & e_{4y} & e_{4z} & -1
\end{vmatrix}
\tag{9.59}
$$

As all $\mathbf{e}_i$ are unity vectors,

$$e_{2x}^2 + e_{2y}^2 = e_{3x}^2 + e_{3y}^2 + e_{3z}^2 = e_{4x}^2 + e_{4y}^2 + e_{4z}^2 = 1 \qquad (9.60)$$

which, together with Equation (9.59), gives

$$\begin{aligned}|\mathbf{G}_u| &= (e_{4y}e_{3z} - e_{3y}e_{4z})(1 - e_{2x}) \\ &+ e_{2y}e_{4z}(1 - e_{3x}) - e_{2y}e_{3z}(1 - e_{4x})\end{aligned} \qquad (9.61)$$

From the definition of GDOP (Equation (9.51)) we have

$$GDOP = \sqrt{\text{trace}[(\mathbf{G}_u^T\mathbf{G}_u)^{-1}]} = \sqrt{\text{trace}[\mathbf{G}_u^{-1}(\mathbf{G}_u^T)^{-1}]}$$

$$= \frac{1}{|\mathbf{G}_u|}\sqrt{\text{trace}[\text{adj}\,(\mathbf{G}_u)\,\text{adj}\,(\mathbf{G}_u^T)]} = \frac{\sqrt{H}}{|\mathbf{G}_u|} \qquad (9.62)$$

where H can be calculated from Equations (9.59) and (9.60), adj is the adjoint matrix and the trace of the matrix is the sum of the diagonal elements. After a tedious calculation we have

$$\begin{aligned}H &= (-e_{2y}e_{3z} - e_{3y}e_{4z} + e_{4y}e_{3z} + e_{2y}e_{4z})^2 + 2(e_{3y}e_{4z} - e_{4y}e_{3z})^2 \\ &+ 2e_{2y}^2(e_{4z}^2 + e_{3z}^2) + (e_{2x}e_{3z} + e_{3x}z_{4z} - e_{4x}e_{3z} - e_{2x}e_{4z})^2 \\ &+ (-e_{3z} - e_{3x}e_{4z} + e_{4z} + e_{4x}e_{3z})^2 \\ &+ e_{4z}^2(e_{2x} - 1)^2 + e_{3z}^2(1 - e_{2x})^2 \\ &+ (-e_{2x}e_{3y} - e_{3x}e_{4y} - e_{4x}e_{2y} + e_{2x}e_{4y} + e_{3x}e_{2y} + e_{4x}e_{3y})^2 \\ &+ (e_{3y} + e_{3x}e_{4y} - e_{4y} - e_{4x}e_{3y})^2 \\ &+ (-e_{2y} - e_{2x}e_{4y} + e_{4y} + e_{4x}e_{2y})^2 \\ &+ (e_{2y} + e_{2x}e_{3y} - e_{3y} - e_{3x}e_{2y})^2 \\ &+ (-e_{2x}e_{3y}e_{4z} - e_{4x}e_{2y}e_{3z} + e_{2x}e_{4y}e_{3z} + e_{3x}e_{2y}e_{4z})^2 \qquad (9.63)\end{aligned}$$

The important conclusion from Equations (9.62)–(9.63) is that the variation in H as a function of position is small because it is a quadratic sum of directional cosines polynomials, and for this reason GDOP is approximately inversely proportional to V. In the first approximation it is therefore optimum to select satellites such that V is as large as possible. A more exact method, however, is to find the minimum of GDOP directly from Equation (9.62). This can be done by minimizing GDOP at the same time as the conditions of Equation (9.60) are taken into account. It is shown in Appendix 8 that

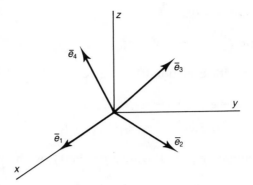

Figure 9.4 Coordinate system for the calculation of GDOP

$(GDOP)_{min}$ obtained in this way is 1.5811. It should always be remembered, however, that the basic assumption behind the derivation of $GDOP$ is that the distance measurement errors are mutually independent and have equal variances, which is very rarely the case.

A time-saving suboptimum method of satellite selection is discussed in Appendix 8.

System descriptions

10 TRANSIT

10.1 Introduction

The history of the TRANSIT system, or NNSS (Navy Navigation Satellite System) as it is also called, started on 4 October 1957 (Stansell, 1978). *Sputnik 1* was launched on that date, and the signals from the satellite were immediately examined very intensely in the United States. At the Applied Physics Laboratory of John Hopkins University, a method was developed to determine the satellite orbit using the Doppler shift of the signals reaching receivers at known positions on the earth. From there it was a short step to invert the method, i.e. to calculate the position of the receiver from the Doppler shift of the signals of satellites in known orbits.

Thus began the development of the first satellite navigation system of the world, TRANSIT, begun early 1959. The system became operative for military use in January 1964 and was released for civil use in July 1967.

The early receivers were expensive and the users not very numerous, but during the last years this has changed rapidly (Figures 10.1 and 10.2).

10.2 The satellites

As long as the system is operative, at any one time it should contain five (possibly more) satellites in polar orbits with an altitude over the earth's surface of 1075 km. The orbit period is about 107 minutes,

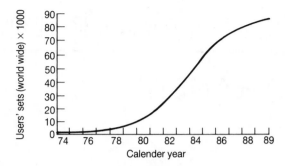

Figure 10.1 The number of TRANSIT receivers in use has increased considerably

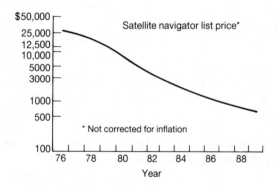

Figure 10.2 Receiver prices have fallen rapidly for many years

implying an orbit velocity of 7.3 km/s. Originally, the orbital planes were evenly distributed around the earth, but they have drifted gradually without significant implications in the use of the system. The orbits of the satellites which were launched up to about 1981 (the 'OSCAR' satellites) could not be controlled after launch, but the newest ones (the NOVA satellites) have possibilites for orbit corrections (Hoskins and Danchik, 1984) (Figure 10.3). There is also a third version, called SOOS, which is a modified OSCAR spacecraft, (launched two at a time on a Scout launch vehicle (Danchik, 1989)), also without orbit maintenance possibilities. At present (Autumn 1989), there are three NOVA spacecraft in orbit, forming the basic constellation (Danchik, 1989).

Because of the small number of satellites and the altitude, the coverage is not time-continuous, but the polar orbits result in a

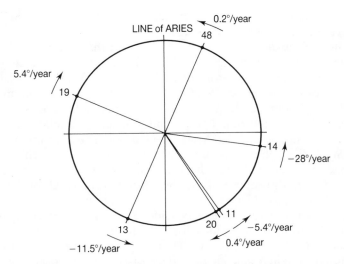

Figure 10.3 Rectascensions of TRANSIT satellites during days 91–92 1984 (Allison and Daly, 1984)

shorter average waiting time as a function of increasing latitudes (Figure 10.4). Usually, only one satellite is visible to the user at a time. Two visible satellites may interfere with each other (depending

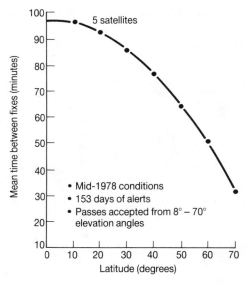

Figure 10.4 Average time between satellite passes 1978 (Stansell, 1978)

on the Doppler shift). Interference from the Soviet TSIKADA satellites (see below) may also occur.

A satellite has a mass of about 61 kg, and the satellite body has a diameter of about 50 cm (Figure 10.5). Four solar cell panels charge the batteries. The power from these is about 30 W in a new satellite, but decreases to about 25 W after five years in space. The antenna, shaped like a lamp shade (Figure 10.5), is directed towards the earth by means of a stabilizing boom, and transmits circularly polarized signals at about 150 and 400 MHz (see below). The transmitted power is 1.5 W–3 W (higher output power from the newer satellites). The antennae (Figure 10.5) must have a relatively wide radiation pattern to cover the earth because of the low altitude (about $\frac{1}{6}$th of the radius of the earth). At the same time, the propagation distance between a satellite and a receiver changes considerably more from low elevation to high elevation than it does for a high-altitude satellite. This has been taken into account at the shaping of the radiation pattern when the satellite antenna gain at the edges of the coverage area is larger than in the middle.

The TRANSIT satellites are controlled by the Navy Astronautics Group headquarters at Point Mugu, California, with control stations in Maine, Minnesota and Hawaii (Stansell, 1978). Each time a satellite is visible from a control station, the station measures and records the signals from the satellite. The results are transferred to the centre at Point Mugu where the satellite orbit is computed and predicted for the following time period. At the subsequent satellite passing of one of the stations, the computed orbital parameters for the following 16 hours are transmitted to the satellite, which stores

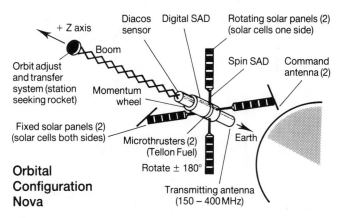

Figure 10.5 TRANSIT satellite (Blanchard, 1983)

them and transmits down to the user. Updating this storage is done approximately twice a day.

10.3 Navigation principles

10.3.1 Satellite signals

A simple block diagram of a TRANSIT satellite is shown by Figure 10.6. Based on a stable frequency reference of about 5 MHz, the satellite transmits two signals at about 150 and 400 MHz. The frequency stability is about 10^{-11} during one satellite pass (i.e. 15–20 minutes). Due to the common reference, the signals are coherent, i.e. the phase of one of them can be derived from the phase of the other. The ratio between the frequencies is 8/3.

Navigational data stored in the satellite are read and phase modulated on to the carrier wave in sequences of two minutes' length with a rate of 6103 bits per 2 minutes, i.e. about 50.9 bits per second. The phase modulation has three levels: $\pm 60°$ and 0 (Figure 10.7). The reason for choosing a modulation form where the phase average is 0 at transmission of '1', as well as at transmission of '0', is that the frequency variation of the received signal on the earth which is used for navigation (the frequency is the derivative of the phase) will not be influenced by the kind of data transmitted and consequently will not influence the position determination. Figure 10.7 also shows that

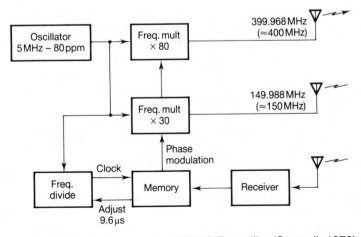

Figure 10.6 Block diagram of a TRANSIT satellite (Stansell, 1978)

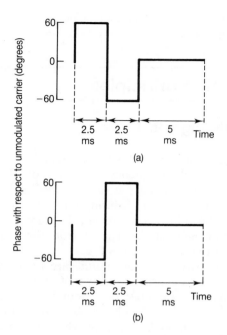

Figure 10.7 Signal format using phase modulation to transfer orbital data from the satellites. '1' is curve (a) followed by curve (b), '0' is curve (b) followed by curve (a)

this data flow furnishes a clock signal at twice the bit-rate. This is utilized to synchronize a clock in the receiver to the data flow. This is important because the data signal is also a synchronizing signal as the transmission of orbital data is controlled so that it starts and stops every even minute. The updating of orbital data from the control centre in California thus includes time corrections as well. These are stored in the memory of the satellite and allowed to adjust the transmitted signals in steps of 9.6 μs (Figure 10.6).

A complete transmission of all navigational data from the satellite to the receiver thus takes precisely 2 minutes. The transferred data are organized in six columns and 26 rows with 39-bit words plus a 19-bit final word (Figure 10.8). The last 25 bits of the whole message form a synchronizing word of 25 bits so that the receiver is prepared for the start of the next 2-minutes' message and synchronized to that.

By this method, the receiver can also find special words in the following message. The message partly contains fixed parameters (the eccentricity of the elliptic orbit, the semi-major axis, etc.), partly

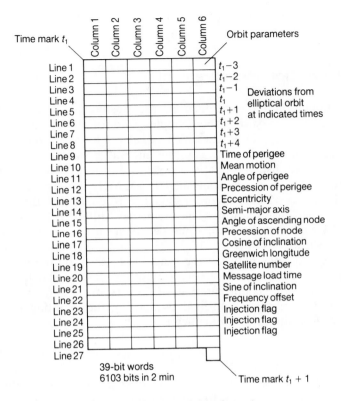

Time mark t_1	Column 1	Column 2	Column 3	Column 4	Column 5	Column 6	Orbit parameters
Line 1							t_1-3
Line 2							t_1-2
Line 3							t_1-1 Deviations from
Line 4							t_1 elliptical orbit
Line 5							t_1+1 at indicated times
Line 6							t_1+2
Line 7							t_1+3
Line 8							t_1+4
Line 9							Time of perigee
Line 10							Mean motion
Line 11							Angle of perigee
Line 12							Precession of perigee
Line 13							Eccentricity
Line 14							Semi-major axis
Line 15							Angle of ascending node
Line 16							Precession of node
Line 17							Cosine of inclination
Line 18							Greenwich longitude
Line 19							Satellite number
Line 20							Message load time
Line 21							Sine of inclination
Line 22							Frequency offset
Line 23							Injection flag
Line 24							Injection flag
Line 25							Injection flag
Line 26							
Line 27							

39-bit words
6103 bits in 2 min   Time mark $t_1 + 1$

Figure 10.8 The data format

parameter corrections (Figure 10.8). The words of rows 1–8 of the message are shifted one step upwards every second minute with a renewal on the eighth row, and by interpolating at the 2 minute intervals the receiver can find the position of the satellite at any time. Rows 9–22 are changed only when a new word enters the satellite memory. Thus, the orbit parameters can be verified by comparison of redundant messages to detect and eliminate occasional errors in the received data.

Only every sixth word is available to a civil user (i.e. the last column of Figure 10.8), and the information starts by word No. 8 and stops by word No. 134 (there are 156 words in the whole message). The last word contains the frequency drift of the satellite and, after 1981, when the first NOVA satellite was launched, also an identification number, distinguishing between older OSCAR and newer NOVA satellites. The remaining $\frac{5}{6}$ of the data in columns 1–5 are used to transfer information about other satellites in the system, among other things.

10.3.2 Position measurements

As mentioned in the introduction, the position of the receiver is determined by measuring the Doppler shift of the signal from the satellite. This Doppler shift, which is caused by the satellite movement relative to the receiver, is in the interval ± 8 kHz.

The receiver usually has a stable frequency reference, f_g, at 400 MHz. The satellites transmit at 400 MHz minus 32 kHz and 150 MHz minus 12 kHz in order that the frequency of the received signal, f_r, is always lower. The frequency difference between the received signal and the local reference then varies during a satellite pass, as shown by Figure 10.9. The number of periods of the difference frequency between the time marks given by the data signal is counted in the receiver. Every row of the data signal lasts about 4.6 s, and a usual count interval of 23 s (some manufacturers use other interval lengths) is formed by starting the counting for every fifth row. The difference frequency is

$$\Delta f = f_g - f_r \tag{10.1}$$

Counting the number of periods of Δf over a certain time interval is the same as integrating the frequency over the same interval, as each count corresponds to a phase increase of 2π. The satellite transmits time marks at t_1, t_2, t_3, ... The counting is started and stopped by these. Even if the satellite transmits the marks at t_1 and t_2, the distance implies that the receiver receives the marks only at $t_1 + \Delta t_1$ and $t_2 + \Delta t_2$ respectively. The content of the counter after the counting interval $t_1 < t < t_2$ is therefore

$$N_1 = \int_{t_1+\Delta t_1}^{t_2+\Delta t_2} (f_g - f_r)\, dt \tag{10.2}$$

The received frequency is equal to the transmitted frequency, f_t, plus the Doppler shift, and is then

$$f_r = f_t(1 + v_r/c) \tag{10.3}$$

where v_r is the velocity component of the satellite towards the receiver, and c is the velocity of light. This gives

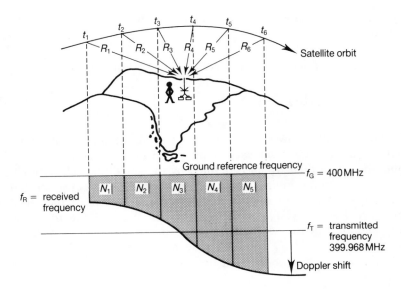

Figure 10.9 The frequency of the received signal is a time variable (Stansell, 1973)

$$N_1 = (f_g - f_t)(t_2 + \Delta t_2 - t_1 - \Delta t_1) - \frac{f_t}{c} \int_{t_1+\Delta t_1}^{t_2+\Delta t_2} v_r \, dt \qquad (10.4)$$

Further, we have (Figure 10.9)

$$R_1 = \Delta t_1 c \qquad (10.5)$$

$$R_2 = \Delta t_2 c \qquad (10.6)$$

where R_1 and R_2 are the distances between the satellite and the receiver at times t_1 and t_2 respectively. The distance change is thus

$$\Delta R = c(\Delta t_1 - \Delta t_2) \qquad (10.7)$$

This distance change is identical to the change given by integration of the velocity difference over the time interval in the receiver, as the time reference comes from the satellite and uses the same time as the rest of the signal on its way down to the receiver.

$$\Delta R = \int_{t_1+\Delta t_1}^{t_2+\Delta t_2} v_r \, dt \qquad (10.8)$$

By inserting Equations (10.7) and (10.8) into Equation (10.4) we obtain

$$N_1 = (f_g - f_t)(t_2 - t_1) + f_g(\Delta t_2 - \Delta t_1)$$

$$= (f_g - f_t)(t_2 - t_1) + \frac{f_g}{c}(R_2 - R_1) \tag{10.9}$$

Equation (10.9) shows clearly what a Doppler count means. Firstly, there is a constant term caused by a constant frequency difference multiplied by a time which is defined by the satellite clock. After that there is a term describing the change in distance between the satellite and the receiver, expressed in number of wavelengths at the receiver reference frequency. (f_g is the scale factor because the time marks are generated in the satellite. If these marks had been generated in the receiver instead, f_t would have been the scale factor.)

In the same way, for the next time interval

$$N_2 = (f_g - f_t)(t_3 - t_2) + \frac{f_g}{c}(R_3 - R_2) \tag{10.10}$$

etc., and for the ith interval,

$$N_i = (f_g - f_t)(t_{i+1} - t_i) + \frac{f_g}{c}(R_{i+1} - R_i) \tag{10.11}$$

The time interval $t_{i+1} - t_i$ is determined in the satellite and known, and the term $R_{i+1} - R_i$ is the only unknown to be solved in each equation. Consequently, the distance difference is measured in exactly the same way as in a terrestrial hyperbolic system. Instead of two transmitting stations at two fixed points, here we have one transmitter moving from one point to the other.

In order to find its own position, the receiver needs to find the intersection point of two hyperbolas. In fact, the surfaces of position of the distance difference are hyperboloids, but these created hyperbolas at the intersection of the plane earth surface. As the position is to be determined in three coordinates, a third measurement interval is needed. This is also the case if, for example, the local reference frequency is unknown.

A satellite at the user's zenith gives a low cross-orbit accuracy; therefore satellites below an elevation angle of about 75° are usually selected. In order to avoid long signal paths through the ionosphere and the troposphere, use of satellites lower than 10–15° is usually avoided. A usable satellite path then lasts for about 10–18 minutes, and with interval lengths of 23 s, typically 20–40 Doppler counts are the result of each satellite pass. This implies that the Equation system (10.11) is overdetermined.

In practice, the position calculation is based on the assumption that the receiver is at a certain position. Knowing the position of the satellite as a function of time, what the Doppler counts would be if the assumed position were correct is computed, and this result is compared to the real count. An iterative procedure (Section 10.6) is used to make the assumed and measured results coincide (analogous to Appendix 5).

Even if the local frequency reference is accurate, it always deviates more or less from 400 MHz. The result is a position error. The errors in the frequency reference of the satellite create corresponding but smaller errors. For this reason the frequency error, $F = f_G - f_t$, is usually treated as an additional unknown in the position computations (the others are the height, the latitude and the longitude, h, Φ and Λ respectively).

10.4 The receiver

Different types of antennae can be utilized for reception, from simple whip antennae to more sophisticated ones. The antenna and the preamplifier are usually co-located. Monopoles (simple or folded) over a ground plane (Figure 10.10) are suitable because they have deep nulls in the radiation pattern in zenith (Figure 10.11), which prevents satellites at high elevation angles from being used. However, 3 dB are lost by linearly polarized antennae as the signal is circularly polarized.

The advantages of the modulation method used are, in addition to minimum frequency jitter, that the clock frequency can be regenerated and there is a possibility of detecting errors as the phase shifts are balanced within each bit. However, the carrier frequency itself carries only about 56 per cent of the transmitted signal power, i.e. about 2.5 dB less. Table 10.1 gives a typical transmission budget (Allison and Daly, 1984). As seen, the low path attenuation of the 150 MHz signal is counterbalanced by the higher external noise level at that frequency.

Figures 10.12 and 10.13 give examples of receiver block diagrams.

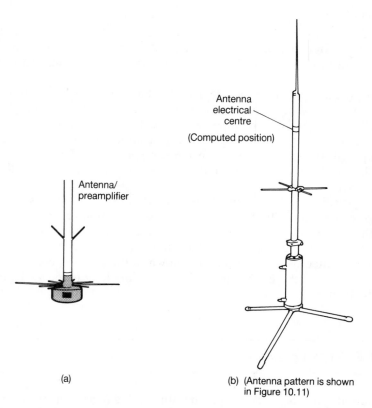

Antenna
electrical
centre
(Computed position)

Antenna/
preamplifier

(a)

(b) (Antenna pattern is shown
in Figure 10.11)

Figure 10.10 Examples of receiver antennae for TRANSIT (Stansell, 1978, 1979)

Figure 10.12 shows the functional blocks of a very common dual-frequency commercial receiver for survey purposes. Fig. 10.13 is an experimental type receiver. (Very little receiver development work has taken place in recent years because of the announced closing of the TRANSIT system in the mid-1990s.)

Usually, there are two intermediate frequencies, but the choice of these is uncritical as long as the second frequency is always higher than the maximum Doppler shift. The LO frequency is, as mentioned, uncritical as long as it is stable, as deviations from a nominal value are taken care of in the position computations. Oven stabilized crystal oscillators are utilized to make the frequency drift low enough during one satellite pass.

An error in the reference frequency (f_g in Equation (10.9)) becomes an error in the measurement of slant range change where

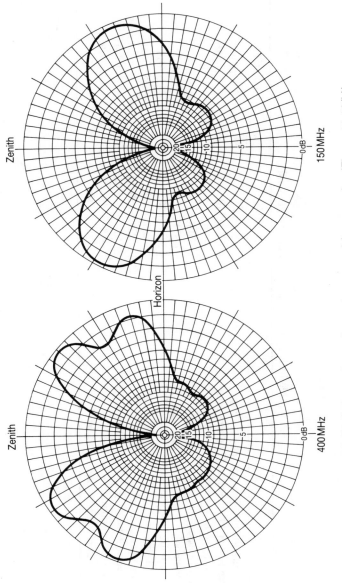

Figure 10.11 Radiation pattern of an antenna used for surveying (Figure 10.10(b))

Table 10.1 Typical link budget of one TRANSIT

	150 MHz	400 MHz
1 Transmit effective isotropic power (eirp)	4.5 dBW	7.0 dBW
2 Space loss at 20° elevation (2290 km range)	−143.2 dB	−151.7 dB
3 Receiver antenna gain, G_r	−1 dB	−1 dB
4 Effective noise temperature at receiver input, T_e	170 K	170 K
5 Antenna noise temperature, T_a	400 K	100 K
6 System noise temperature, $T_s = T_e = T_a$	27.6 dBK	24.3 dBK
7 Ground station, G_r/T_s	−28.6 dB/K	−25.3 dB/K
8 Boltzmann's constant	−228.6 dB((W/K)/Hz)	
9 Receiver $C/N = (1 + 2 + 7 − 8)$ in dB-Hz	61.3 dB-Hz	58.6 dB-Hz
10 Total received power, $C = (1 + 2 + 3)$	−139.7 dBW	−145.7 dBW
11 Total received carrier power, $C_r = C − 2.5$ db	−142.2 dBW	−148.2 dBW

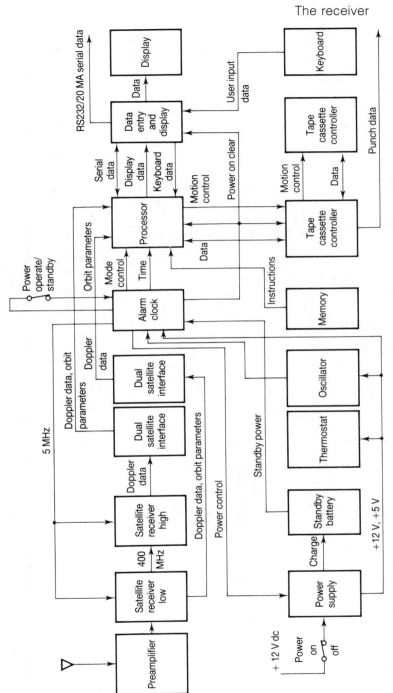

Figure 10.12 Functional block diagram of the MX1502 survey receiver (Stansell, 1978. Courtesy Magnavox Advanced Prod. Div.)

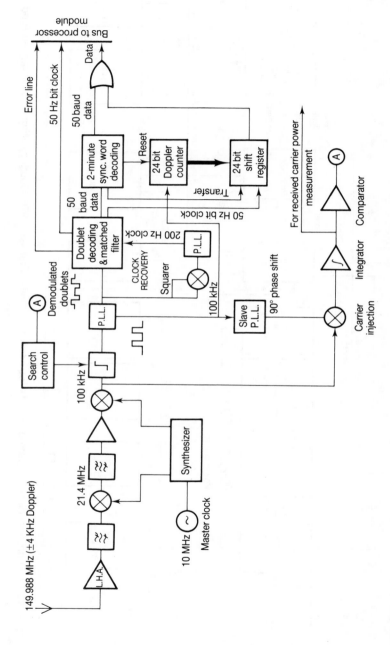

Figure 10.13 An example of a TRANSIT receiver block diagram (the 150 and 400 MHz channels are principally equal) (Allison and Daly, 1984)

each count is usually equivalent to about one metre or less. Jitter in the time recovery process from the satellite signal also contributes to a Doppler count error. A high-precision receiver clock can replace the timing from the satellite signal (i.e. by defining interval limits t_{i+1}, and t_i in Equation (10.11), whereby f_g is replaced by f_t).

In the example shown in Figure 10.13, the signal is split after the 2nd IF stage, and the carrier is coherently demodulated in one signal branch to give the received power level. In the other branch the phase is demodulated (after a limiter) to obtain the navigational data. Two phase-locked loops (of second order) are used. The bandwidth of the main loop, about 10–20 Hz, determines the noise bandwidth of the receiver. This bandwidth also discriminates against signals from other TRANSIT satellites over the horizon, as these have other Doppler shifts. (A certain interference from navigation satellites in the Soviet TSIKADA system may occur as these transmit at 150.000 and 400.000 MHz with tone signalling. The tones at 3.5 and 7.0 kHz are amplitude modulated on to the carrier transmitting data with 50 baud. There are twelve satellites in this system (Daly and Perry, 1986)).

When the search for the satellites starts with no knowledge of the position ('cold start'), the VCO frequency has to be swept over the whole possible frequency range (i.e. ±8 kHz), but more often the receiver can be fed with an approximate position, so that the preknowledge enables it to search only within a limited frequency range. The sweep is stopped when a signal is detected. The control functions are taken care of after detection of the signals from the second intermediate frequency stage. The received signal level has to exceed a predetermined value before the signal is accepted. Noise pulses during a satellite pass may bring the receiver out of lock, but positioning can still be made if at least a number of Doppler counts equal to the number of unknowns to be solved are made before the interruption, or if the receiver relocks to the same satellite having enough time to do the necessary number of counts. In both cases, however, the accuracy suffers because of the reduced number of counts.

The positioning itself, implying a comparison of the measured Doppler curve with computations at assumed positions, only takes a few seconds after the conclusion of the counting. The presentation of the result varies depending on receiver type, but the resolution is often only a few metres. In addition, the time of the measurement is also displayed. The role of the operator is usually modest, often only keying in an initial position with an accuracy of the order of 100 n.m. and GMT within a quarter of an hour.

10.5 The use of two carrier frequencies

On its way from the satellite to the receiver, the signal has to pass through the ionosphere and the troposphere. The propagation velocity in the troposphere depends on pressure, temperature and the relative moisture of the air, but relatively accurate empirical formulas exist for this dependence (Equation (3.26)). By measurement of these parameters and use of the values for correction, the remaining position errors caused by the troposphere, can be made very small in most cases. Variations in refractivity along the signal path also have to be taken into account. It should be noted that the moisture of the air decreases much more rapidly with height than with pressure. Most receivers use standard atmosphere values to correct for tropospheric delays.

The reduction of the influence of the ionosphere requires measurements at two carrier frequencies, however, and the use of a two-channel receiver so that the delay at the two frequencies can be compared. (The influence of the troposphere on propagation velocity, on the other hand, is frequency-independent.) The increased phase velocity of the signal in the ionosphere thus gives an increased effective wavelength (because $v_f = f\lambda$, Figure 10.14) and thus also a seemingly more curved satellite orbit.

This virtual change of the satellite orbit is caused by the Doppler shift, which can be written $f_d = v_r/\lambda$. This is most influenced when

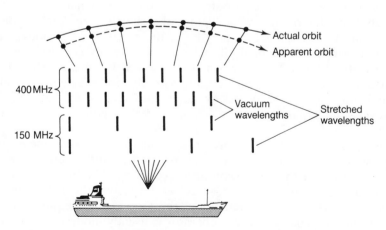

Figure 10.14 Passing through the ionisphere gives a larger effective wavelength (Stansell, 1978)

the signal has the longest path through the ionosphere, i.e. at low elevation angles. This gives the same influence as if the satellite orbit formed a larger angle with the line-of-sight (i.e. smaller v_r at constant λ), i.e. a more curved orbit.

The Doppler shift is

$$f_d = f_t \frac{v_r}{v_f} \tag{10.12}$$

where v_f is equal to the phase velocity ($v_f \approx c$) and, consequently,

$$\Delta f_d = f_t \frac{-v_r}{c^2} \Delta v_f \tag{10.13}$$

The change in time delay of the signal when passing the ionosphere is inversely proportional to the square of the frequency (Equation (3.2)), i.e. $\Delta v_f \sim f_t^{-2}$ and then

$$\Delta f_d = K/f_t \tag{10.14}$$

The constant of proportionality, K, is dependent on time and place, but a coarse estimation, based on empirical data, can be performed. Most TRANSIT receivers (about 90 per cent) only receive the 400 MHz signal, and only a coarse correction of the ionospheric influence can then be performed. If, however, the receiver receives both the 150 and the 400 MHz signals, Equation (10.14) can be utilized. The Doppler counts of the two frequencies are then assumed to be

$$N'_{150} = N_{150} + \frac{K}{150} \tag{10.15}$$

$$N'_{400} = N_{400} + \frac{K}{400} \tag{10.16}$$

where N' denotes the actual result of the counting, and N the result without the ionospheric influence. But

$$N_{400} = \tfrac{8}{3} N_{150} \tag{10.17}$$

and Equations (10.15)–(10.17) together give

$$N_{400} = \tfrac{24}{55} \left(\tfrac{8}{3} N'_{400} - N'_{150} \right) \tag{10.18}$$

or

$$(\Delta N_{400})_{\text{corr.}} = N_{400} - N'_{400} = \tfrac{24}{55}(\tfrac{3}{8}N'_{400} - N'_{150}) \qquad (10.19)$$

where Equation (10.19) gives the correction of the measured counting result. (It might be added that other delay differences between the 400 and 150 MHz signals, in the satellite or in the receiver, may reduce the accuracy of the above method.)

10.6 The iteration procedure

The position computations are based on the method of least squares (Appendix 5), and proceed in the following way: the difference between the ith measurement result at the actual position and the corresponding computed measurement result at the assumed position is denoted e_i. This is a function of positioning error and frequency error. The difference must be minimized, and the wanted quantities are therefore the latitude $(\Delta\Phi)$, the longitude $(\Delta\Lambda)$ and the frequency difference $(\Delta F = \Delta f_{\mathrm{g}} - \Delta f_{\mathrm{t}})$ changes as functions of the assumed position. (In the case of 3-D positioning, the height must also be included.) The difference between the measurement and the computed result, e_i, obtains the new value e'_i when the parameters are changed where

$$e'_i = e_i - \frac{\partial e_i}{\partial \Phi}\,\Delta\Phi - \frac{\partial e_i}{\partial \Lambda}\,\Delta\Lambda - \frac{\partial e_i}{\partial F}\,\Delta F \qquad (10.20)$$

where $e_i = O_i - C_i$ (Equation (A5.5)).

But when the set of Equations (10.20) is overdetermined, i.e. $i = 1, 2, \ldots, n$, where n lies between 20 and 40, the aim of the parameter change is not to minimize the single e_i, but to minimize the sum of squares

$$s = \sum_{i=1}^{n} e_i^2 \qquad (10.21)$$

On matrix form the set of equations can be written

$$\mathbf{PD} = \mathbf{E} \qquad (10.22)$$

where

$$\mathbf{P} = \begin{bmatrix} \dfrac{\partial e_1}{\partial \Phi} & \dfrac{\partial e_1}{\partial \Lambda} & \dfrac{\partial e_1}{\partial F} \\[2ex] \dfrac{\partial e_2}{\partial \Phi} & \dfrac{\partial e_2}{\partial \Lambda} & \dfrac{\partial e_2}{\partial F} \\[1ex] \vdots & & \\[1ex] \dfrac{\partial e_n}{\partial \Phi} & \dfrac{\partial e_n}{\partial \Lambda} & \dfrac{\partial e_n}{\partial F} \end{bmatrix} \qquad (10.23)$$

$$\mathbf{D} = (\Delta\Phi \ \ \Delta\Lambda \ \ \Delta F)^{\mathrm{T}} \qquad (10.24)$$

$$\mathbf{E} = (e_1 \ e_2, \ldots, e_n)^{\mathrm{T}} \qquad (10.25)$$

This is mathematically identical to the procedure described in Appendix 5 and the solution of Equation (10.22) is consequently

$$\mathbf{D} = (\mathbf{P}^{\mathrm{T}}\mathbf{P})^{-1}\mathbf{P}^{\mathrm{T}}\mathbf{E} \qquad (10.26)$$

When the three parameter changes $\Delta\Phi$, $\Delta\Lambda$, ΔF have been solved according to Equation (10.26), they are added to the original position estimate, and thus new values of e_i and its derivatives can be computed. Subsequently, new values of D are computed according to Equation (10.26), etc. Normally only two or three iterations are needed, even if the first assumed position is several tens of kilometres from the actual one. The method leads to convergence, even with such large errors of the assumed position as 200–300 km but, in this case, several iterations may be needed. If the third coordinate (the altitude) of the receiver is unknown, the same method as given above is utilized, in that case with more Doppler counts.

10.7 Accuracy

Positioning errors principally have two essentially different causes: (i) errors in the measurements themselves caused by receiver velocity (see below), inaccuracies in the receiver and the satellite instrumentation, noise, propagation variations, etc.; and (ii) errors in the translation of measurement results into position, i.e. inaccurate computations. The latter type of errors can, in principle, be reduced to an arbitrarily small quantity just by iterating a sufficient number of times.

The most important sources of error for a stationary receiver and their influence on position accuracy are the following (rms errors in metres) (Piscane *et al.*, 1973):

- uncorrected residual errors from tropospheric effects   1–5
- uncorrected residual errors from ionospheric effects
  (two-channel receiver)   1–5
- noise (thermal noise, transmitter and receiver phase
  jitter, clock noise)   3–6
- different delays in the two receiver channels   4–9
- uncertainty in the altitude of the receiving antenna
  (determination of vertical position gives another
  unknown in Equations (10.20)–(10.26), erroneous
  altitude estimation in the computation of horizontal
  position gives the largest error longitudinally because
  of the direction of movement of the satellite)   10
- satellite orbit uncertainty in the extrapolation interval
  because of gravity variation, air drag and radiation
  pressure (applies for OSCAR, NOVA has about the
  half)

along orbit:	10–25
cross orbit:	3–6
vertically:	6–12

- datum shift   2–10

rms error of one satellite pass   18–35 m

A one-channel receiver gives a considerable increase in the first item of the above error budget, so that the total error is in the range 80–100 m. This implies that, for a stationary receiver, uncorrected tropospheric and ionospheric errors give the largest single contribution to the final result. The two-channel receiver, on the other hand, where most of the propagation errors have been eliminated, gets its largest error contribution from uncertainties in the satellite orbit. The reason is that the orbit has to be predicted. (Post-processing of measurement results from the monitor stations can determine the orbit with about 5 m uncertainty. The use of such orbital data gives a considerable improvement in accuracy.)

Utilization of more satellite passes results in considerable improve-

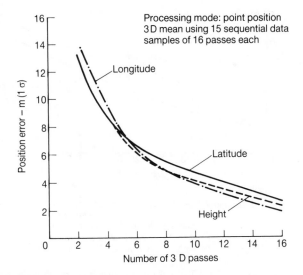

Figure 10.15 Improvement of position accuracy by utilization of more satellite passes (Blanchard, 1983)

ments of position accuracies because of reduced variance in the final measurement error. Figure 10.15 shows what can be achieved. Generally, position accuracy is higher in latitude than in longitude because of the polar orbits of the satellites.

If the receiver is not stationary but moves with an unknown velocity, the error is larger than that of a stationary receiver. This is due to the fact that the result of the Doppler counting then corresponds to the average position in the interval and the fact that the utilization of the redundancy of the measurements ($n > 3$ in Equation (10.20)) is reduced. In such cases the receiver has to try to estimate its own velocity and to utilize the estimated velocity as a correction factor. Figure 10.16 shows position errors as a function of velocity errors. (This is rms value, the maximum error may be 2–5 times larger.)

10.8 Translocation

When a stationary two-channel receiver is used, the uncertainty of the satellite orbit is, as already stated, the largest source of error. This uncertainty can be reduced considerably by post-determination of the

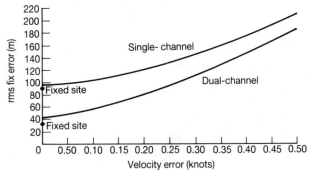

Figure 10.16 rms position error as a function of erroneous estimation of the receiver's own velocity. This applies to an arbitrary direction. If the direction is known, the error is smaller (Stansell, 1978)

orbit using measurement results from monitor stations instead of pre-determination by means of extrapolation. The method, however, requires positioning data from the receiver to be recorded and processed only when the post-determined accurate orbital data have been received from the monitor station. Such a method which can reduce the position error of the two-channel receiver to well below 10 m requires a stationary receiver and plenty of time, in addition to the accurate orbital data from the monitor stations.

A better method is usually the so-called method of translocation, which gives as good, or better results on essentially the same basis. It is based on the fact that the largest part of the error when using TRANSIT is caused by factors outside the receiver. Two receivers tracking the same satellite at the same time and the same place show very similar measurement results, i.e. the errors are strongly correlated. Results show that this correlation is present even when the receivers are separated, and the separation distance may be as large as 200 km. Thus a receiver can determine its position in relation to another receiver with an accuracy of 1 m, or sometimes better, over considerable distances. The method implies that recorded measurement errors at one base station at a known position are utilized to correct the measurement results at the receiver whose relative position relative to that of the former is to be determined (cf. differential OMEGA).

In Norway (Blankenburgh and Coldevin, 1976) and in other countries a considerable number of translocation measurements have been performed. Measurement data from the two stations have then been post-processed. The results show that about fifty satellite passes is optimum. The method is frequently used when positioning platforms where accuracy requirements are 5 m or better.

10.9 Fields of utilization/Future

TRANSIT has had a most important application in the precise determination of stationary positions, e.g. in survey, oil prospecting and drilling, hydrographic measurement and surveillance. The system is used on board many types of ships as well, to the extent that costs and accuracies are improved, e.g. on fishing boats and surface marine ships, but also on pleasure boats. As indicated by the name NNSS, TRANSIT is very much used by the navy. Submarines utilize it when surfacing to update their inertial systems.

As TRANSIT does not allow continuous measurements to be performed, with regard to navigation it can only be used for updating of other aids, e.g. dead reckoning (integrating the velocity vector), Doppler radar or inertial systems. These aids must have limited drift between satellite passes.

According to today's plans (Autumn, 1989) TRANSIT will be phased out from the beginning of 1997, as the military needs can then be met by means of GPS (Danchik, 1989).

11 Spread spectrum: coding of satellite signals

11.1 Introduction

When using satellite systems based on distance measurements, the transmitted signal has to be marked with a time reference, so that the receiver is capable of measuring the time of arrival of the signal in the time reference system used by the satellites. Most forms of modulation enable time marks to be read, but the precision and ease of implementation differ. The modulation frequency is determined by the requirements for distance resolution. Using digital modulation as an example, if each bit is regarded as a pulse with the same length, the bit-length, τ, can be calculated approximately from Equation (4.5), when the permitted time error Δt and the signal-to-noise ratio are given. If, for example, the distance error permitted is 1 m (i.e. $\Delta t = 3$ ns) and the signal-to-noise ratio is 20 dB, $\tau \approx \sqrt{100}\ 3\pi/2$ ns ≈ 50 ns. Usually the distance measurements are carried out in such a way that the times of arrival of several bits are averaged (sampled), while the bit-length is taken into consideration, so that the signal-to-noise ratio (in the resulting bandwidth) obtains a higher value due to the integration than that which is valid for each bit; but the approximate calculation shows the order of magnitude of the necessary bit-length/bit-rate. When the modulation frequency has been determined by the accuracy requirements in the navigation system, the modulation type has to be determined. For some reason it may be desirable that all satellites in the system utilize the same carrier frequency. In that case the type of modulation must contain certain characteristics making it possible to separate the different satellites from each other with regard to the signals. Military systems also require that the signal be difficult to detect and disrupt.

This is where a technique called *spread spectrum* can be utilized. It was developed mainly for transmission of digitized speech and data, but development in the field of high-frequency digital technology has contributed to a considerable increase in its use. A name which is used frequently in communications is code-division multiple access (CDMA). The modulation types, characteristic of each satellite, are also called codes. It should be added, however, that this technique is used also in systems with one code common to many transmitters, e.g. the Soviet GLONASS (Chapter 13).

It is the use of spread spectrum in GPS in particular which has provoked the interest of the navigation community in this method, but it should be mentioned that it has also been used in ground systems for some time, e.g. in Syledis (Laurent and Nard, 1975). The method is also used in a few other terrestrial positioning systems, e.g. in Geoloc (Nard, 1984), and it is planned to be used in the European Space Agency's satellite system NAVSAT (Joyce *et al.*, 1989).

A description of the principles of spread-spectrum techniques used in satellite navigation systems is given below in order to facilitate understanding of the functions of such systems.

11.2 Spread spectrum

11.2.1 Principles

The name *spread spectrum* (Harris, 1973) is due to the fact that the power of the signal to be transmitted by means of extra modulation is spread over a bandwidth many times larger than the information bandwidth (Figure 11.1). (By 'information' in this respect, we mean the navigation message, e.g. 50 bits/s in GPS. This terminology, which is inherited from communication systems, does not reflect the fact that there is indispensable information in the spread-spectrum signal itself in navigation systems.) The signals $g(t)$ in Figure 11.1 thus have a bandwidth, B_g, much larger than the bandwidth, B_m, of the information signal.

In the receiver the spreading function $g_1(t)$ is known, and the signal processing in the receiver starts with a compression of the received signal by correlating with the locally generated or stored $g_1(t)$, i.e. the receiver carries out the operation

$$v_1(t) = \int_{-\infty}^{\infty} \left[\sum_{i=1}^{n} g_i(t - t_i) S_i(t - t_i) + n_i(t) \right] g_1(t - t_1)\,dt \qquad (11.1)$$

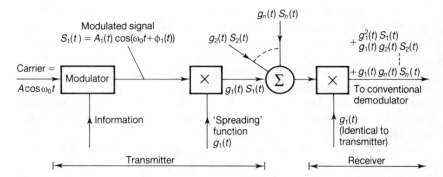

Figure 11.1 Principles of a spread-spectrum system. (The original version of Figure 11.1 through 11.10 was first published by the Advisory Group for Aerospace Research and Development, North Atlantic Treaty Organisation (AGARD/NATO) in Lecture Series LS 58 – *Spread Spectrum Communications* in July 1973)

Note that the time of integration is very large (cf. the integration limits in Equations (11.1) and (11.2)) here only when the variations in g_i are considered. On the other hand, S_i has a small bandwidth in relation to g_i and is to be considered as a constant at the integration. Now orthonormal spreading functions $g_i(t)$ have been selected, so that, ideally,

$$\int_{-\infty}^{\infty} g_i(t - t_i)g_j(t - t_j)\,\mathrm{d}t = \begin{cases} 1 \text{ if } i = j \\ 0 \text{ otherwise} \end{cases} \tag{11.2}$$

This means that when $g_1(t)$ is selected as the despreading function in the receiver, $S_1(t)$ is passed to signal processing in the conventional demodulator; when $g_2(t)$ is selected, $S_2(t)$ is demodulated, etc. This immediately shows one of the great advantages of spread spectrum: the same carrier frequency can be utilized for more than one channel.

11.2.2 Spreading functions

The requirement for spreading functions $g_i(t)$ can be separated into two parts:

$$R_{ii}(\Delta t) = \int_{-\infty}^{\infty} g_i(t)g_i(t + \Delta t)\,\mathrm{d}t = \begin{cases} 1 \text{ if } \Delta t = 0 \\ 0 \text{ otherwise} \end{cases} \tag{11.3}$$

and

$$R_{ij}(\Delta t) = \int_{-\infty}^{\infty} g_i(t)g_j(t + \Delta t)\, dt = 0 \qquad (11.4)$$

for any Δt when $i \neq j$. Equation (11.3) expresses the autocorrelation properties of the function; $R_{ii}(\Delta t)$ is the autocorrelation function, whereas Equation (11.4) expresses the cross-correlation properties; $R_{ij}(\Delta t)$ is the cross-correlation function for $g_i(t)$ and $g_j(t)$. The functions $g_i(t)$ and $g_j(t)$ are orthogonal if Equation (11.4) is satisfied.

The choice of the spreading functions g_i is far from trivial and, in practical cases, usually

$$R_{ii}(\Delta t) = \begin{cases} 1 & \text{if } \Delta t = 0 \\ \ll 1 & \text{if } |\Delta t| > \delta \end{cases} \qquad (11.5)$$

where δ is a correlation distance. The requirement in Equation (11.4) is often still more difficult to meet. Depending on the system specifications normally a limit is set such that

$$|R_{ij}(\Delta t)| < \varepsilon,\, 0 < \varepsilon \ll 1, \qquad i \neq j \qquad (11.6)$$

The modulation of the carrier can, in principle, be achieved in different ways. The modulation may be AM, FM, PM or combinations of these. In order that the spread power be as uniformly distributed as possible in the frequency band, and it is important to ensure that the power density and, consequently, the probability of interference (in both directions), is low, it is necessary that the modulation have a noise character. The most suitable types of modulation are biphase modulation with PN codes (PN = pseudo noise), frequency hopping between a great number of (seemingly) randomly chosen carrier frequencies and time hopping. The latter type of modulation implies pulse position modulation where the pulse interval is varied according to a noise-like pattern.

Bi-phase modulation (Figure 11.2) is preferred for navigation systems where the bit transitions are to be utilized to measure the arrival time of the signal. At the same time modulation, as well as demodulation, of the carrier with low rate information should be relatively simple.

All $g_i(t)$ are sequences of 0 and 1 where one symbol corresponds to the phase 0 and the other to the phase π (Figure 11.2). The method is particularly suitable when the low rate information ($\Phi_1(t)$

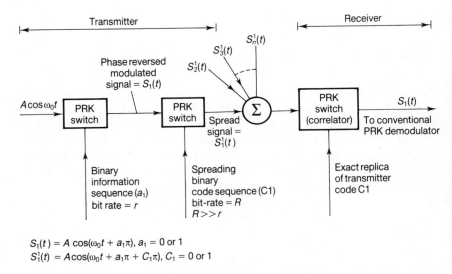

$S_1(t) = A\cos(\omega_0 t + a_1\pi)$, $a_1 = 0$ or 1
$S_1^1(t) = A\cos(\omega_0 t + a_1\pi + C_1\pi)$, $C_1 = 0$ or 1

Figure 11.2 Spread-spectrum system with biphase modulation

in Figure 11.1) is also presented in binary form. This can simply be superimposed on the spreading function by means of exclusive-or addition. (Exclusive-or addition, written $\oplus$, means that $1 \oplus 1 = 0 \oplus 0 = 0$ and $0 \oplus 1 = 1 \oplus 0 = 1$.) The bit-rate of the spreading sequence is much larger than that of the information sequence. Figure 11.3 exemplifies how the phase of the carrier varies as a function of the binary modulation, and it shows what happens in the receiver at the demodulation, partly with the desired signal, and partly with an undesired signal of the same type.

As the bi-phase modulation is the method utilized in navigation satellites, other modulation principles, e.g. frequency hopping and time hopping, are not treated here.

11.2.3 The correlation function

The choice of binary sequence or code depends on the requirements for autocorrelation and cross-correlation properties. It is desired to have an equal number of 0s and 1s in the phase of the transmitted carrier wave to make the DC component of the modulation signal 0. The expected value of the autocorrelation would then be exactly 0 for Δt larger than one bit-length (Equation (11.3)), if the bit sequence was random. Figures 11.4(a) and (b) show samples of such a code and the calculation of its autocorrelation function. The autocorrela-

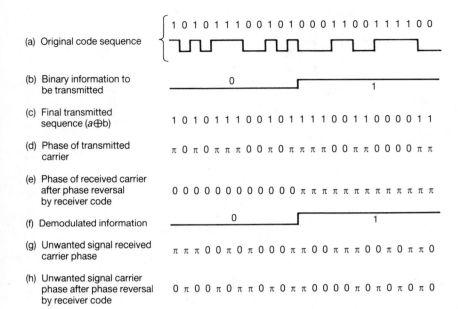

(a) Original code sequence

(b) Binary information to be transmitted

(c) Final transmitted sequence ($a \oplus b$)

(d) Phase of transmitted carrier

(e) Phase of received carrier after phase reversal by receiver code

(f) Demodulated information

(g) Unwanted signal received carrier phase

(h) Unwanted signal carrier phase after phase reversal by receiver code

Figure 11.3 Example of signal format and the influence of the receiver correlator on wanted and unwanted signals

tion of a function $f(t)$ is generally defined as

$$R(\tau) = \lim_{T \to \infty} \left[\frac{1}{2T} \int_{-T}^{T} f(t) f^*(t + \tau)\, dt \right]$$
(11.7)

With the notation of the diagram we have

$$R(\tau) = A^2 (1 - |\tau|/T_1)$$
(11.8)

The power spectrum of the signal is the Fourier transform of the autocorrelation function, and the power spectrum of a random binary code is then

$$S(\omega) = A^2 \int_{-\infty}^{\infty} (1 - |\tau|/T_1) \exp(-j\omega\tau)\, d\tau$$

$$= A^2 \int_{-T_1}^{0} \left(1 + \frac{\tau}{T_1}\right) \exp(-j\omega\tau)\, d\tau$$

$$+ A^2 \int_{0}^{T_1} \left(1 - \frac{\tau}{T_1}\right) \exp(-j\omega\tau)\, d\tau$$

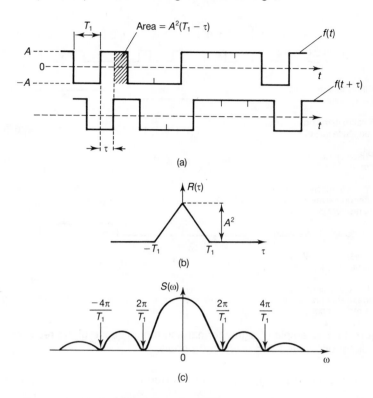

Figure 11.4 The autocorrelation function (**b**) and power spectrum (**c**) of a random, binary code (**a**)

$$= A^2 \int_{-T_1}^{0} \left[\frac{-1}{j\omega} \left(1 + \frac{\tau}{T_1} \right) + \frac{1}{\omega^2 T_1} \right] \exp(-j\omega\tau)$$

$$+ A^2 \int_{0}^{T_1} \left[\frac{-1}{j\omega} \left(1 - \frac{\tau}{T_1} \right) - \frac{1}{\omega^2 T_1} \right] \exp(-j\omega\tau)$$

$$= \frac{A^2}{\omega^2 T_1} \left\{ 2 - \left[\exp(j\omega T_1) + \exp(-j\omega T_1) \right] \right\}$$

$$= \frac{2A^2}{\omega^2 T_1} (1 - \cos \omega T_1)$$

$$= \frac{4A^2}{\omega^2 T_1} \sin^2 \left(\frac{\omega T_1}{2} \right) = A^2 T_1 \left(\frac{\sin \dfrac{\omega T_1}{2}}{\dfrac{\omega T_1}{2}} \right)^2 \tag{11.9}$$

$S(\omega)$ is shown in Figure 11.4(c).

Figure 11.5 indicates what happens to an unwanted signal. During the correlation process in the receiver the bandwidth of the desired signal is reduced down to the information bandwidth, whereas the other signals are modulated by the spreading function and their powers are distributed over a corresponding frequency range. The integration of Equations (11.3) and (11.4) is carried out by filtering in a bandpass or lowpass filter where the bandwidth corresponds to the information bandwidth B_m. Because the powers of the unwanted signals are distributed over a wide frequency band by the multiplication in the correlator, a smaller part of those powers passes through the filter. The result is that the ratio between the signal and the interference is improved by approximately the factor B_g/B_m, the so-called processing gain. This also applies to the interferences at one signal frequency. White noise, on the other hand, which has constant power density independent of frequency, is uncorrelated with the spreading function and therefore has about the same power density after multiplication. This also applies to other undesired signals with the same spectrum as the desired one but with other spreading functions.

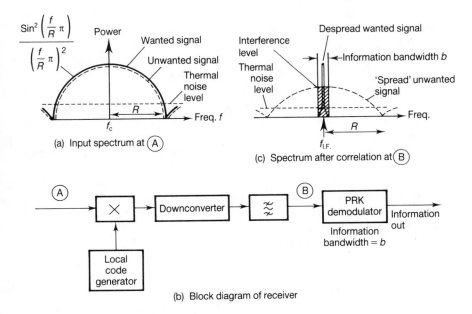

(a) Input spectrum at (A)

(c) Spectrum after correlation at (B)

(b) Block diagram of receiver

Figure 11.5 Spectra before and after correlation

11.2.4 Generation of the codes (spreading functions)

As mentioned above, it may be desirable to have a random sequence of 0s and 1s in the transmitted code because the expected autocorrelation function is then exactly zero outside the correlation peak. However, it is impossible to utilize such a code sequence if this should be used by the receiver at the same time. For this reason the code has to be deterministic, even if it looks random to an observer over a certain time interval. The term 'pseudo random' is used. A large group of such codes is generated in feedback shift registers, and because of the feedback the code is repeated after a certain period length. Figure 11.6 shows the principles of such code generators. If the register content in Figure 11.6(b) is 0001 at the beginning, the process is as shown by Table 11.1.

For each clock pulse the register content is shifted one step to the right, and the content of the first cell is the exclusive-or (modulo-2) sum of the previous content of the third and fourth cells. (Linear and non-linear sequences are generally differentiated, which means that linear and non-linear operations, respectively, are allowed for generation. A binary code being linear means that only modulo-2 addition is allowed. Only linear codes are treated here as these are the only codes used in navigation systems.)

It is common to describe such linear code generators by means of polynomials of the form $1 + \sum x^i$, where x^i means that the output of the ith cell is used as input to the mod-2 adder, and 1 means that the output of the adder is fed to the first cell. The code generator in Figure 11.6(b) is thus described by $1 + x^3 + x^4$.

As shown by the example of the table, the sequence is repeated after $p = 2^N - 1$ bits, where N is the length of the register. Because of the feedback the state of zeros only in the register should not occur as this state is a stable one, so for this reason the period cannot be 2^N, which is the maximum number of states of a register of length N.

Such shift register sequences are called PN sequences (Pseudo Noise) of maximum length (m sequences). There are feedback loops which do not create PN sequences of maximum length. Table 11.2 shows the number of possibilities of generating PN sequences of maximum length at different register lengths.

Because the number of 1s is always 1 larger than the number of 0s in such sequences, the autocorrelation function is not exactly 0 outside the correlation peak but has the value $-1/p$ where p is the number of bits of the sequence. As the sequence is repeated, the spectrum is not continuous but is a line spectrum where the line

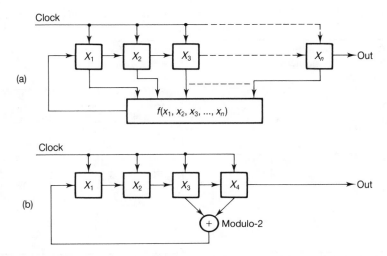

Figure 11.6 Code generator with shift register: **(a)** generalized; **(b)** of 4th order (linear)

Table 11.1 PN sequence of a shift register with four cells

x_1	x_2	x_3	x_4	Out
0	0	0	1	1
1	0	0	0	0
0	1	0	0	0
0	0	1	0	0
1	0	0	1	1
1	1	0	0	0
0	1	1	0	0
1	0	1	1	1
0	1	0	1	1
1	0	1	0	0
1	1	0	1	1
1	1	1	0	0
1	1	1	1	1
0	1	1	1	1
0	0	1	1	1
0	0	0	1	1

separation is $1/pT$ (Figure 11.7). Therefore, the power spectrum can be calculated as a Fourier series expansion from Figure 11.7(a), and the coefficients are

Table 11.2 The number of possible m sequences for shift register lengths up to 21

Shift register length	Total number of sequences	Sequence length	Shift register length	Total number of sequences	Sequence length
2	1	3	12	144	4095
3	2	7	13	630	8191
4	2	15	14	756	16 383
5	6	31	15	1800	32 767
6	6	63	16	2048	65 535
7	18	127	17	7710	131 071
8	16	255	18	7776	262 143
9	48	511	19	27 594	524 287
10	60	1023	20	24 000	1 048 575
11	176	2047	21	84 672	2 097 151

Note: Image sequences (where every 0 and 1 have been interchanged) are also included.

Source: Harris, 1973.

$$c_n = \frac{1}{pT} \left\{ \int_{-T}^{T} \left[1 - \frac{|\tau|}{T} \left(1 + \frac{1}{p} \right) \right] \exp\left(-jn\omega\tau\right) d\tau \right.$$
$$\left. - \frac{1}{p} \left(\int_{-p(T/2)}^{-T} \exp\left(-jn\omega\tau\right) d\tau + \int_{T}^{p(T/2)} \exp\left(-jn\omega\tau\right) d\tau \right) \right\}$$

(11.10)

where $\omega = (2\pi/pT)$ and $n = 0, \pm 1, \pm 2, \ldots$ Equation (11.10) gives

$$c_n = \frac{(1 + p)\sin^2\left(n\dfrac{\omega T}{2}\right)}{p^2 \left(n\dfrac{\omega T}{2}\right)^2} - \frac{\sin(n\pi)}{np\pi}$$

$$= \frac{1 + p}{p^2} \cdot \frac{\sin^2\left(\dfrac{n\pi}{p}\right)}{\left(\dfrac{n\pi}{p}\right)^2} - \frac{\sin(n\pi)}{pn\pi}$$

(11.11)

Equation (11.11) shows that the envelope of the spectrum components of the repeated shift register sequence of maximum length has

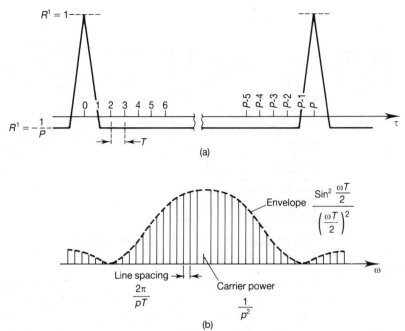

Figure 11.7 Normalized autocorrelation function of a PN sequence (**a**), and its power spectrum; (**b**)

the same shape as the spectrum of the ideal sequence, except for the carrier frequency (Figure 11.7(b)).

11.2.5 The receiver

Figure 11.8 shows a generalized block diagram of a spread-spectrum receiver. It contains one loop to track the carrier and another to track the code. Code tracking means synchronizing the receiver-generated code to that received from the transmitter. The two most common methods are shown in Figures 11.9–11.10.

A delay-lock loop implies that two correlators with mutually displaced phases are utilized (Figure 11.9(a)). The outputs of the two correlators are subtracted, and when the difference is zero, the two locally generated code phases are symmetrical on each side of the correlation peak (Figure 11.9(b)). (It is assumed that at least one of the channels is within less than one bit from the correlation peak.) The subtraction can also be carried out before the correlation as this

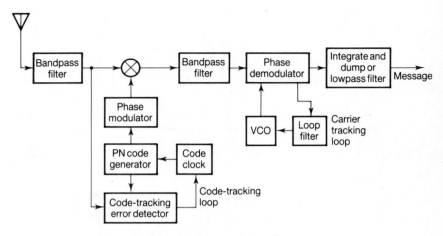

Figure 11.8 Spread-spectrum biphase demodulator

is a linear process (Figure 11.9(c)). In order to hit the correlation peak exactly, which gives maximum signal-to-noise ratio, the delay has to be Δ (Figure 11.9(a)), in which case a third correlator is needed. The three channels are sometimes called *early*, *late* and *on-time*.

The position of the correlation peak is used to define the time-of-arrival of the satellite signal with reference to the local clock. Thus, the accuracy of the code tracking directly influences the ranging accuracy. The accuracy is determined by the S/N in the code tracking loop, in full analogy with phase measurements, as described in Section 4.2.

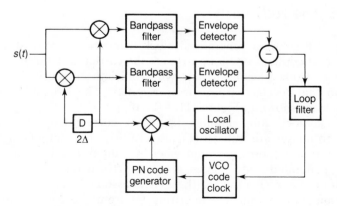

Figure 11.9(a) Principle of delay-lock loop

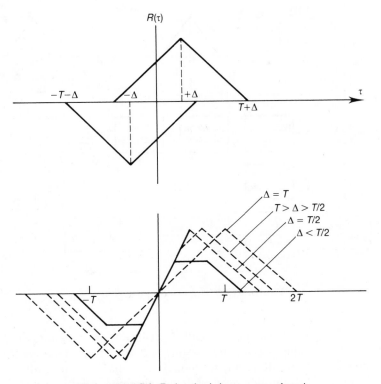

Figure 11.9(b) Delay-lock loop error signal

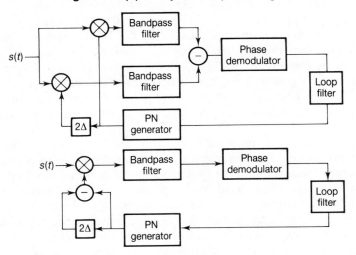

Figure 11.9(c) Alternative methods for delay-lock loop

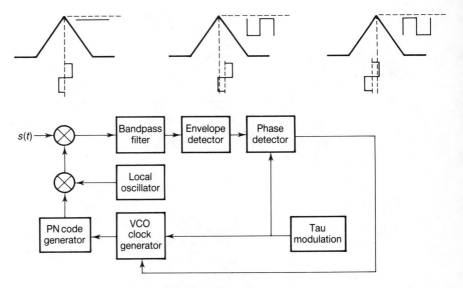

Figure 11.10 τ-jitter code tracking

Another method is the, so-called, τ-jitter or τ-dither method
(Figure 11.10). This implies modulating the local code delay by a low
frequency. This modulation frequency can be detected after the
correlator as this implies a modulated climbing on the correlation
function. The phase of this detected signal depends on which side of
the peak a local code phase is (Figure 11.10). The output signal of
the phase detector is utilized to control the local phase so that the
modulation lies on the peak. τ-jitter does not require as many
components (only one correlator) as the delay-lock loop, but it does
give a lower signal-to-noise ratio due to the fact that the local code
phase always fluctuates between two points on each side of exact
synchronism and, consequently, somewhat down from the correlation
peak.

Modern receivers incorporate digital mechanizations using the
above principles. The correlators can be realized by in-phase and
quadrature sampling of IF or baseband signals and modulo-2 addition
of the corresponding local code chips. This is followed by an
accumulator reconstructing the autocorrelation function as the
number of agreements minus the number of disagreements. Code
tracking is implemented by using a much higher clock frequency
(sampling rate) than the code chip rate (i.e. high resolution, e.g. $\frac{1}{256}$
of a code chip), early and late channels as an up- and down-counting

accumulator estimating the code phase, and the output of the accumulator controlling the phase of the code NCO (numerically controlled oscillator). In order to remove the effect of data transitions, the sign of the code phase (i.e. the accumulator output) is inverted or not according to the sign from the on-line correlator in the I channel.

It should be noted that a matched filter is mathematically equivalent to a correlator. Both mechanizations carry out the operations of Equation (11.1). Such a filter may consist of a shift register of the length of one or several code periods. Each cell is tapped, and the output is weighted in such a way that the weights form an image of the code (Dixon, 1976).

Besides the fact that the receiver-generated code has to be locked to the received code, the carrier wave of the received signal has to be regenerated. Regeneration of the carrier is usually made in the receiver at an intermediate frequency (in principle, the frequency has no significance as long as it is much larger than the bit-rate).

The reason for the need to regenerate the carrier is that this is part of the data demodulation. Figure 11.2 and Equation (11.1) show that the signal consists of data modulated on to a carrier if the spread-spectrum modulation is removed. Contrary to the codes the navigation message is not known, so it has to be found by coherent demodulation, i.e. by means of a signal in phase with the received carrier. Thus it is necessary to regenerate the carrier for demodulation of the data signal.

One of the methods for carrier regeneration in bi-phase modulation is squaring of the signal. This implies that

$$v(t) = \{a \cos[\omega t + \Phi(t)]\}^2 = \frac{a^2}{2}(1 + \cos 2\omega t) \qquad (11.12)$$

where $\Phi(t) = 0$ or π. The phase modulation is removed in the process described by Equation (11.12). The phase-locked loop that is utilized can also be a Costas loop (Figure 11.11), which in principle gives the same result. As shown by Figure 11.11, the regenerated carrier in the loop can be found at the VCO output (VCO = voltage controlled oscillator), and the phase modulation is found at the output from the lowpass filters in the side branches. The filter determining the loop characteristics, i.e. the filter with the narrowest bandwidth, is the one immediately in front of the VCO in Figure 11.11.

When the carrier frequency is a fixed multiple of the code frequency, only one controlled oscillator is necessary in the receiver.

Carrier tracking in digital receivers is implemented by digital

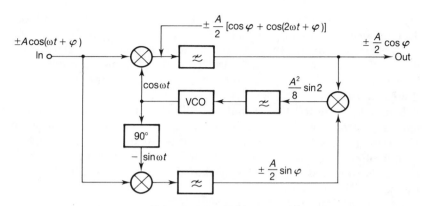

Figure 11.11 Costas loop

phase-lock loops, very similar to the digital code loop described above. An estimate of the incoming carrier phase is usually derived as arctan (Q/I) where Q and I are the quadrature and in-phase signal samples respectively. The calculation involves a look-up table of the arctan function. Before phase-lock is achieved, the resulting phase rotates with regard to the locally generated I and Q signals. Bit transitions between phase estimates must be removed, as in the analog Costas loop, and this can be done using the sign from the on-time correlator as mentioned above for the code phase estimate. The phase estimates are filtered in a digital filter before applying corrections to the carrier NCO.

The receiver functions are then, in principle, as shown in Figure 11.12. A carrier loop regenerates the carrier and a code loop tracks the received codes. In Figure 11.12 code tracking is done by means of a delay-lock loop (early–late correlator, see Figure 11.9(a)), and the locally generated code is modulated onto an LO signal which is not phase-locked to the received signal. Another possibility is to modulate the PN code onto the signal of the VCO in the carrier loop.

Synchronization of the system, i.e. locking of the loops, then occurs in the following way. The phase of the locally generated code is displaced continuously or stepwise until it coincides with the one received, within one bit-length. The correlator then gives an output signal, whose amplitude is proportional to the position of the correlation function (Figure 11.4(b)). This output signal, whose frequency is the difference between the frequency of the input signal and that of the VCO in the carrier loop, makes this loop capable of locking (without correlation between a local and a received code the input signal of the carrier loop consists of noise only, which after passing

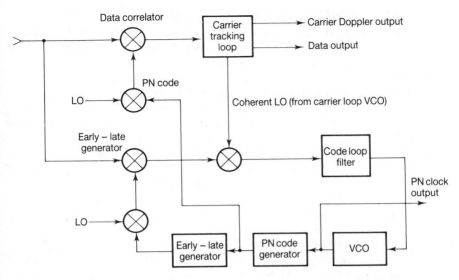

Figure 11.12 Principle block diagram of a spread-spectrum receiver for biphase PN modulation. (The original version of Figure 11.12 was first published by the Advisory Group for Aerospace Research and Development, North Atlantic Treaty Organisation (AGARD/NATO) in Lecture Series LS 58 – *Spread Spectrum Communications* in July 1973)

the loop filter cannot make the loop lock). Thus the code loop locks first and the carrier loop immediately after that.

11.3 Spread spectrum in NAVSTAR/GPS

Spread-spectrum modulation by means of PN codes is particularly attractive when arrival times are to be measured accurately. The reason for this is that the code loop tracks the received code with a time error which necessarily has to be a fraction of a bit-length. The code-tracking error increases if the phase transitions are not distinct, because the correlation peak then has a more rounded shape. Noise, of course, makes it more difficult to keep good synchronism between a locally generated and a received code. With distinct phase transitions and a good signal-to-noise ratio there are no particular difficulties in keeping the synchronization error in the order of 0.01 bit-lengths.

In NAVSTAR GPS two types of code are used, the P-code and the C/A-code. P means precision or protected, and the bit-length of this code is $\frac{1}{10.23}$ μs. In principle, the code is a compound PN code with a period of about 267 days, but out of this only 7 days for each satellite are utilized, after which the code is reset (see below). The term 'protected' indicates that this code is reserved for certain (mainly military) uses. Due to the length of the total code it is not possible to utilize the code if the user does not know the code as a function of time, because the search procedure then takes too long. In addition, the code is encrypted (see below).

The other code is for use by civil users and also as an aid to the P-code users for locking to the P-code. It is called the C/A-code (C/A = clear/aquisition). Its name is derived from these two functions. The code length is 1023 bits. It runs at the rate of 1.023 Mbit/s, and the repetition frequency is thus 1 kHz.

The C/A-code is generated as shown in Figure 11.13 (Spilker, 1978). Two shift registers, each having ten cells, generate one PN code each of length 1023 bits. These codes are added in an exclu-

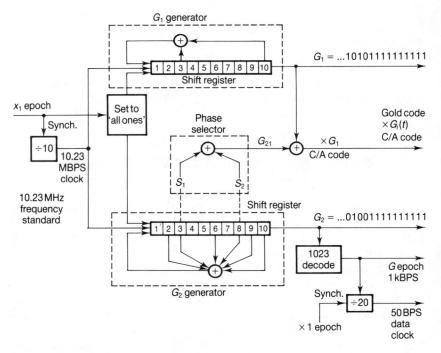

Figure 11.13 Generation of the C/A code (Spilker, 1978)

sive-or adder, and the mutual phase shift is determined by the couplers S_1 and S_2. The reason is that a PN-code has the easily verifiable property that, added to a phase shifted version of itself, it does not change but just obtains another phase. The task of S_1 and S_2 in the diagram is thus to shift the codes G_1 and G_2 in relation to each other to enable the generator to form of a great number of codes. The polynomials are $G_1 = 1 + X^3 + X^{10}$, and $G_2 = 1 + X^2 + X^3 + X^6 + X^8 + X^9 + X^{10}$.

The reason why shift register sequences of maximum length are not utilized directly is partly due to the fact that the number of available codes is relatively small (thirty, because for practical reasons image codes cannot be utilized, see Table 11.2), and partly because the cross-correlation properties are not as good as is desirable. The autocorrelation function has sidelobes, too, if the integration time is short (i.e. a small number of code lengths). The shorter this time, the greater the probability of high sidelobes leading to a locking onto a wrong correlation peak.

Codes generated as in Figure 11.13 are called *Gold codes* (Gold, 1967). Two certain shift register sequences can be used to generate $2^n + 1$ Gold codes as each sequence can be shifted in $2^n - 1$ steps in relation to the other one, and in addition, each of the two sequences can be regarded as a Gold code. Out of the thirty existing maximum length sequences with $n = 10$, $\frac{30!}{2!28!} = 435$ combinations of two by two codes can be selected. As each such pair can generate 1023 Gold codes, the total number of possible Gold codes is $435 \times 1023 + 30 = 445,035$. Among these codes there is a large probability of finding a sufficient number with good cross-correlation properties.

The cross-correlation between Gold codes for code lengths of 1023 bits is $< -21.6 \, \mathrm{dB}$, irrespective of the Doppler shift of the signal (Spilker, 1978).

The autocorrelation function and the spectrum of the Gold codes (Figure 11.14) deviate a little from those of PN sequences, but this is not very significant.

The NAVSTAR P-code is also generated by PN sequences (Yakos *et al.*, 1975), but in that case the registers have 12 cells (Figure 11.15). Each PN sequence then has the length $2^{12} - 1 = 4095$ bits. The polynomials are: $X1A = 1 + X^6 + X^8 + X^{11} + X^{12}$, $X1B = 1 + X^1 + X^2 + X^5 + X^8 + X^9 + X^{10} + X^{11} + X^{12}$, $X2A = 1 + X^1 + X^3 + X^4 + X^5 + X^7 + X^8 + X^9 + X^{10} + X^{11} + X^{12}$, and $X2B = X^2 + X^3 + X^4 + X^8 + X^9 + X^{12}$ (NATO, 1987).

The registers X1A and X2A are reset after 4092 bits and the registers X1B and X2B after 4093 bits so that the sequences are truncated. When X1A has run through 3750 such cycles, which means

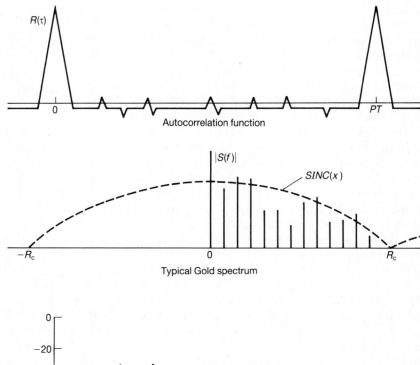

Figure 11.14 Autocorrelation function and spectrum of Gold codes (Spilker, 1978)

$3750 \times 4092 = 15,345,000$ bits, X1A and X1B are both reset. Immediately before this, X1B is kept stationary for 343 bit-lengths after having run through 3749 cycles with 4093 bits in each cycle. X2A and X2B are controlled in the same way as X1A and X1B, respectively, but with the difference that when 15,345,000 bits have been run

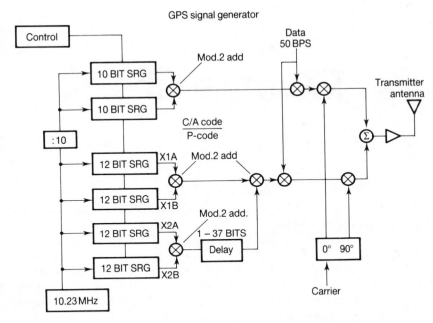

Figure 11.15 Code generation in NAVSTAR satellites (Yakos, 1975)

through, which takes exactly 1.5 s, X2A and X2B are stationary for 37 bit-lengths. The X2 registers together, therefore, have a period of 15,345,037 bits. The output P-code, which is the exclusive-or sum of X1 and X2, thus has the length $15,345,000 \times 15,345,037 = 2.35 \times 10^{14}$ bits. With 10.23 Mbits/s such a sequence has a period of 266.4 days. However, each satellite uses only 7 days of the whole sequence, and restarts again at Saturday midnight. The different satellites use different parts of the total sequence, and with $21 + 3$ satellites, $21 \times 7/266.4 = 63$ per cent of the whole available format is utilized.

In the Block-II satellites the P-code is encrypted, and the encryption code is added to the P-code described above so that the receiver needs a code key to be able to utilize the information. This is the normal mode of operation and is termed anti-spoofing (A-S), which indicates that the purpose is to prevent code transmission by unfriendly forces. The encrypted P-code is denoted as the Y-code (NATO, 1987). The decryption is accomplished in one A-S module per hardware channel within the receiver. A-S techniques and characteristics are classified.

$\mathbb{12}$ NAVSTAR/GPS

12.1 Introduction

NAVSTAR (Navigation System with Time and Ranging) is a satellite system under construction. Its history starts at the end of the 1960s with experimental satellites launched by the US Navy to investigate whether time measurements referred to accurate clocks in satellites could replace and improve TRANSIT frequency measurement (Easton, 1978). The first of these experimental satellites was *Timation I* (low orbit satellite), which was launched in 1967. Another satellite of a similar type, *Timation II*, launched in 1969, was then followed by the *Navigation Technology Satellite* (NTS), *1*, *2* and *3* in 1974, 1977 and 1981 respectively. The last two were the first NAVSTAR satellites, having orbits of about 20,200 km altitude, and inclinations of about 63°. The Timation satellites had quartz clocks on board and transmitted at VHF/UHF, whereas *NTS 1*, *2* and *3* were equipped with much improved clocks as technology progressed, quartz and rubidium, quartz and cesium, and cesium and hydrogen maser, respectively. This improvement is illustrated by the relative clock drift per day which for the five satellites was 300, 100, 5–10, 1–2 and 0.1×10^{-13}, respectively (Bartholemew, 1978). The so-called Block-II satellites of GPS should contain four clocks: 2 Rb and 2 Cs. The clock drift will then be 2×10^{-13}, i.e. one second in 158,000 years.

12.2 Satellite orbits

The GPS satellites will use orbits at an altitude of 20,200 km. The

272

orbits are circular (eccentricity ≪ 0.03), and the inclination is 55° (applies to the Block-II satellites). The constellation will be as shown by Figure 12.1. There are six orbital planes, evenly spaced around the equator, with four satellites in each, planned for when the system is operational some time in the middle of the 1990s. This is called the 21-satellite primary constellation (Green *et al.*, 1989), which means that there are twenty-four satellite positions, but because of a certain

	(Classical coordinates, km/deg)					
	a	*e*	*i*	Ω	ω	*M*
A1	26 609.0	0.	55.0	325.730 284	0.	190.96
A2	26 609.0	0.	55.0	325.730 284	0.	220.48
A3	26 609.0	0.	55.0	325.730 284	0.	330.17
A4	26 609.0	0.	55.0	325.730 284	0.	83.58
B1	26 609.0	0.	55.0	25.7302 839	0.	249.90
B2	26 609.0	0.	55.0	25.7302 839	0.	352.12
B3	26 609.0	0.	55.0	25.7302 839	0.	25.25
B4	26 609.0	0.	55.0	25.7302 839	0.	124.10
C1	26 609.0	0.	55.0	85.7302 839	0.	286.20
C2	26 609.0	0.	55.0	85.7302 839	0.	48.94
C3	26 609.0	0.	55.0	85.7302 839	0.	155.08
C4	26 609.0	0.	55.0	85.7302 839	0.	183.71
D1	26 609.0	0.	55.0	145.730 284	0.	312.30
D2	26 609.0	0.	55.0	145.730 284	0.	340.93
D3	26 609.0	0.	55.0	145.730 284	0.	87.06
D4	26 609.0	0.	55.0	145.730 284	0.	209.81
E1	26 609.0	0.	55.0	205.730 284	0.	11.90
E2	26 609.0	0.	55.0	205.730 284	0.	110.76
E3	26 609.0	0.	55.0	205.730 284	0.	143.88
E4	26 609.0	0.	55.0	205.730 284	0.	246.11
F1	26 609.0	0.	55.0	265.730 284	0.	52.42
F2	26 609.0	0.	55.0	265.730 284	0.	165.83
F3	26 609.0	0.	55.0	265.730 284	0.	275.52
F4	26 609.0	0.	55.0	265.730 284	0.	305.04

Epoch = 1989. 11. 26. 0. 0. 0.0

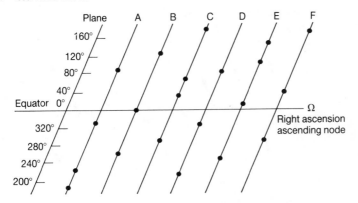

Figure 12.1 The 21 primary satellite constellation (Green *et al.*, 1989)

failure probability, the number of usable satellites will vary between twenty-one and twenty-four. There will always be at least four satellites above 5° elevation, with PDOD < 10 for each user.

The satellites in each orbit will not be evenly spaced in relation to each other, as shown by Figure 12.1, and neither is the displacement from one plane to the other (see the table in Figure 12.1). The fourth satellite in each orbital plane has a position which makes rapid repositioning possible if one of the others fails.

Originally, GPS was to obtain twenty-four satellites in three orbital

	(Classical coordinates, km/deg)					
	a	e	i	Ω	ω	M
A1	26 609.0	0.	55.0	325.730 284	0.	190.882 752
A2	26 609.0	0.	55.0	325.730 284	0.	329.882 752
A3	26 609.0	0.	55.0	325.730 284	0.	87.132 752 3
B1	26 609.0	0.	55.0	25.7302 839	0.	260.882 752
B2	26 609.0	0.	55.0	25.7032 839	0.	358.882 752
B3	26 609.0	0.	55.0	25.7302 839	0.	129.882 752
C1	26 609.0	0.	55.0	85.7302 839	0.	289.882 752
C2	26 609.0	0.	55.0	85.7302 839	0.	68.882 752 3
C3	26 609.0	0.	55.0	85.7302 839	0.	172.632 752
D1	26 609.0	0.	55.0	145.730 284	0.	328.132 752
D2	26 609.0	0.	55.0	145.730 284	0.	86.6 327 523
D3	26 609.0	0.	55.0	145.730 284	0.	216.882 752
E1	26 609.0	0.	55.0	205.730 284	0.	12.132 7523
E2	26 609.0	0.	55.0	205.730 284	0.	108.882 752
E3	26 609.0	0.	55.0	205.730 284	0.	247.632 752
F1	26 609.0	0.	55.0	265.730 284	0.	42.882 752 3
F2	26 609.0	0.	55.0	265.730 284	0.	173.132 752
F3	26 609.0	0.	55.0	265.730 284	0.	291.632 752
A4	26 609.0	0.	55.0	325.730 284	0.	224.382 752
E4	26 609.0	0.	55.0	205.730 284	0.	150.882 752
C4	26 609.0	0.	55.0	85.7302 839	0.	35.382 752 3

Epoch = 1989. 11. 26. 0. 0. 0.0

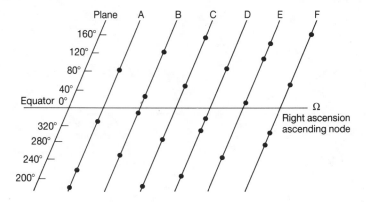

Figure 12.2 The 21 optimal satellite constellation (Green *et al.*, 1989)

planes with eight in each (Milliken and Zoller, 1978). After some years of the programme, the number was reduced to eighteen (in six orbital planes) for economical reasons in 1980, but this appeared to give imperfect coverage in certain areas during short periods. The system was supplemented by three so-called active (i.e. transmitting) reserves. Even this appeared to give unacceptable values of PDOP in certain areas for shorter periods of time (up to about half an hour) (NATO, 1987; Johannesen, 1987–8). However, only those who needed 3-D navigation, i.e. air traffic, would suffer, and those areas would move with time because of the orbital time of the satellites (see below). The disadvantage was still considered serious, so it was decided to expand the final constellation to twenty-four. Integrity aspects also played a part in this decision.

However, there will be an interim arrangement as a so-called optimal 21-satellite constellation should be maintained until the final 24-satellite constellation is operative. This interim constellation is shown in Figure 12.2. and implies adjustments of the satellite positions of the earlier planned constellation with 18 + 3 (Green *et al.*, 1989).

The orbital times of the satellites so far have been, accurately, 12 sidereal hours, i.e. 11 h 58 min. This means, among other things, that each satellite follows the same track, but with about 4 min time delay per 24 hours – an advantage from a test point of view, but this also means that the orbit is in synchronism with the earth's rotation, which, because of the asymmetric mass distribution of the earth, causes resonance in the satellite orbit. This in its turn implies that the satellite orbit must be corrected about once a year to be kept within tolerances. Such a measure takes about three days for each satellite, and during this time it cannot be used for its purpose.

The satellites (Block-I) which were used in the test period had an inclination of 63° and the position was optimized to give the best coverage of the test range at Yuma, Arizona. An inclination of 63.45° of a satellite orbit prevents the orbit from rotating around an axis through the earth's centre perpendicular to the orbit plane (i.e. a drifting of the argument of perigee) caused by the flattening of the earth. This is, however, less important in the case of a circular orbit. The inclination of 55° was chosen for two reasons: (i) the space shuttle was to be utilized for the launches; and (ii) two satellite orbits in opposite directions should intersect at a 90° angle (Figure 12.3), which means good geometry to the user. The first reason is, however, no longer valid because of the delays of the GPS program which followed the space shuttle disaster in January 1986 and which also caused a decision to use the *Delta-II* rocket for launching.

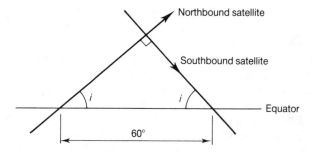

Figure 12.3 North and south proceeding satellite orbits intersect at 90° angle (Equation (A2.6), Appendix A2.1, gives $\operatorname{ctan}^2 = \cos 60°$, i.e. $i = 55°$)

12.3 The satellites

Each satellite weighs 1667 kg (Sherod, 1989) and has a design lifetime of about 7.5 years (tests so far have shown that a 10 year lifespan can be expected). It has two solar cell panels which are able to produce around 600 W and NiCd-batteries for eclipse power generation. The operational satellites are three-axes stabilized by means of infrared detectors (directed towards the earth) and angle gyroes (during the stabilization period).

The antennae are helix array. The polarization is right-handed circular. The largest gain is 15 dB. For navigational purposes two carriers are transmitted (to enable the user to correct for ionospheric delays, Section 3.1) at the frequencies $L_1 = 1575.42$ MHz and $L_2 = 1227.6$ MHz. The carriers are coherent (i.e. the phase of one carrier can be derived from the phase of the other), and the frequency ratio is 154/120. The carriers are biphase modulated in quadrature with a data signal at 50 bits/s and the spread-spectrum signal (Figure 11.15). L_1 contains both the codes and the signal, L_2 only the P-code and the data signal. (It is also technically possible to transmit the C/A-code on L_2.)

Table 12.1. shows a link budget for the navigational signals.

The satellites are controlled from ground stations by means of coded and encrypted signals at S-band. The ground control uses the up-link frequency 1783.74 MHz and the down-link frequency 2227.5 MHz. These signals are not coherent with navigational signals, and have been designed to be particularly jam-resistant.

Table 12.1 Link budget of user navigation signals

	Zenith			5° elevation angle		
	L_1		L_2	L_1		L_2
	C/A	P	P	C/A	P	P
Satellite transmitter power (dBW)	14.25	11.25	6.35	14.25	11.25	6.40
Cable and antenna losses (dB)	1.0	1.0	1.0	1.0	1.0	1.0
Polarization losses in the antenna (dB)	0.25	0.25	0.25	0.25	0.25	0.25
Antenna gain (dB)	15	15	15	12	12	12
Effective radiated power (dBW)	28	25	20.1	25	22	17.15
Space loss (dB)	182.5	182.5	180.6	184.2	184.2	182.3
Atmospheric loss (dB)	0	0	0	0.85	0.85	0.85
Received power on ground (0 dB antenna gain) (dBW)	−154.5	−157.5	−160.5	−160	−163	−166

12.4 The control segment

GPS consists of four main parts (Figure 12.4): the space segment (i.e. the satellites); the user segment (i.e. the navigation receivers); the control segment (i.e. the segment for management and control of the satellite functions and signals); and the segment for detection and reporting of nuclear explosions. From the navigational point of view only the first three are of interest. The control segment (Francisco, 1984) consists of three parts: the master control station; the monitor stations; and the up-link ground antennas. The geographical positions are shown in Figure 12.5.

The master control station is the operations centre for the whole GPS and is located in Colorado Springs (Consolidated Space Operations Center). It contains personnel and equipment for all manage-

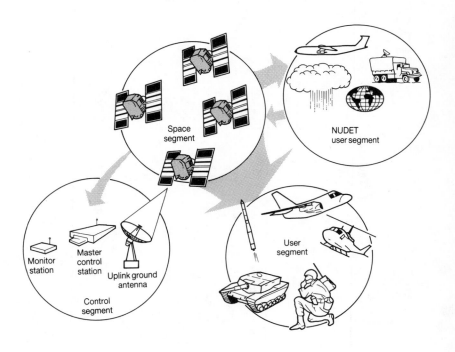

Figure 12.4 The main parts of GPS (Francisco, 1984)

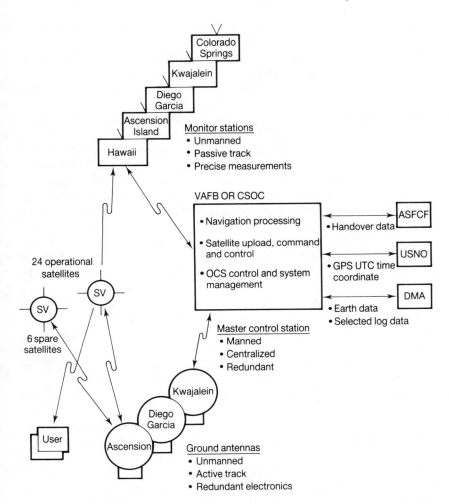

Figure 12.5 The GPS control segment (Francisco, 1984)

ment control of the system. All processing and computation of data from the monitor stations needed to keep the satellites in their orbits, together with their defined functions, is carried out in this centre. The navigation message to be transmitted from the satellites down to the users is also produced here. The centre has constant contact with the satellites by means of the up-link ground antennae, surveying and ensuring that everything works according to plan and that all parameters are kept within tolerances. Communications to and from the monitor and ground antenna stations use satellites other than GPS.

The control centre must also keep GPS time within UTC ± 180 ns (95 per cent) plus possible leap seconds. The deviation from UTC should be included in the navigation message.

The monitor stations are located as shown in Figure 12.5. Their task is to track the satellites, check that all signals meet the requirements and to measure the ranges to the satellites. The latter is done by means of particularly accurate navigational receivers and antennae covering all satellites above 5° elevation and which have particularly low sensitivity to environmental reflections (14 dB discrimination of reflections when the satellite is above 15° elevation). The receivers at the monitor stations are controlled by cesium clocks and track all visible satellites at the same time (multichannel receiver). The ranges, as well as the range rates, are tracked, and measured data are filtered by means of special algorithms and references to the receiver's own known position and to the demodulated navigation messages such that, for example, clock drift in the satellites can be surveyed.

The up-link ground antennae, also shown in Figure 12.5, are unmanned, like the monitor stations. They communicate with the satellites at the mentioned S-band frequencies, receiving data about satellite functions, up-linking control signals and up-dating the content of the navigational messages. These messages are produced in the master control station based on data from the monitor stations. This is based on the use of models for the satellite movement under the influence of external forces such as solar, earth and lunar gravitation, radiation pressure from the sun and earth and particle-induced drag. Even relativistic quantities are part of the computations. The datum used is WGS-84 (Section 1.2). Raw data from the monitor stations are analysed thoroughly and compared with threshold values (this also applies to the first and second derivatives). Smoothing with a time constant of 1.5 s is used; in this way low-noise data are produced every 15 min. The smoothing is performed by a Kalman filter with ten state variables (the satellite position, velocity and acceleration, the phase, frequency and ageing of the satellite clock, the sun radiation intensity, the phase and frequency of the clock of the monitor station, and the height of the troposphere at the monitor station). Navigational data are up-linked to the satellite every 8 h, and the parameters of the navigational message to the users are then matched to the computational results of the centre according to the method of least squares. (It should be mentioned that Block-II replenishment satellites to be launched after 1992 will have a cross-link capability, making intersatellite transmissions possible, so that all up-link information can be transmitted directly from the master control station.)

12.5 The navigation message

12.5.1 Introduction

The navigation message in GPS (NATO, 1987; van Dierendonck *et al.*, 1978) is the information received by the users from the satellites. This is a 50-bit/s data sequence, bi-phase modulated on both carriers, in the same way as the spread-spectrum codes (Figure 11.15). The message has a total length of twenty-five frames of 1500 bits each, and each frame is subdivided into five subframes of 300 bits each (ten words of 30 bits each). Thus each subframe is transferred in 6 s, each frame in 30 s and the whole message in 12.5 min. Each subframe begins with a telemetry word (TLM) and a hand-over word (HOW). The details of the contents of the different subframes are given below. The format is shown in Figures 12.6 and 12.7.

12.5.2 TLM and HOW

The telemetry word consists of an 8-bit preamble for synchronizing, a 14-bit message about the state of the satellite and the message itself, 2 empty bits and 6 parity bits. The hand-over word contains a time-of-week count of 17 bits, which gives the number of X1 epochs since the beginning of the present week, one bit of synchronization control, another control bit, 3 bits for identification of the subframe, 2 empty bits and 6 parity bits. The TOW-count is utilized by the receiver at the transition from the C/A- to the P-code (hence the name, 'hand-over word'), which is necessary if a P-code receiver does not originally know its own position in relation to the satellite within 10–20 km. The Z-count is the time stated as the X1 epochs of the P-code at the start of the next subframe and is a 29-bit number whose 19 least significant bits represent the TOW-count. The 10 most significant bits represent the week number since 5 January 1980.

12.5.3 Data block I

This data block occupies the last 240 bits of the first subframe (Figure 12.6) and is repeated every 30 s. It is generated by the control

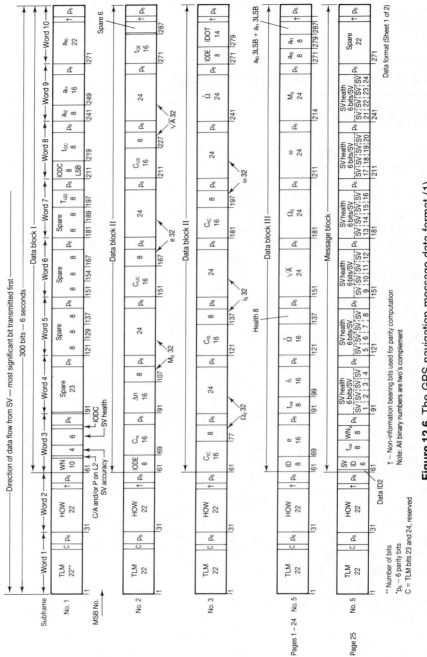

Figure 12.6 The GPS navigation message data format (1)

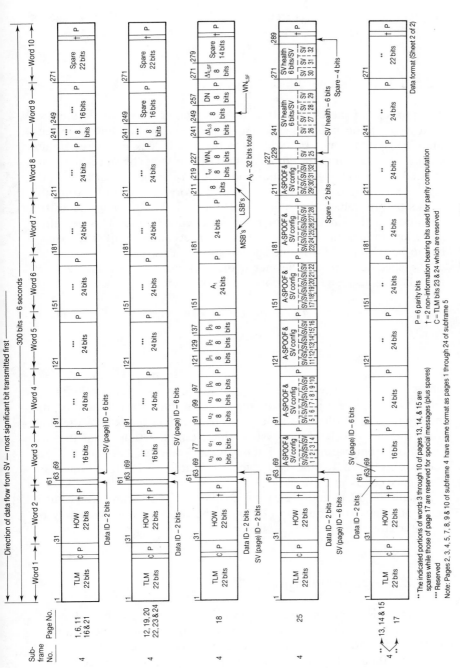

Figure 12.7 The GPS navigation message data format (2)

Data format (Sheet 2 of 2)

segment and contains corrections to the satellite clock and age of transmitted data.

12.5.3.1 Clock corrections

The purpose of the clock correction parameters is to provide the user with information about the deviation between the satellite clock and the GPS system time. In addition to the usual clock drift the satellite clocks are also influenced by relativistic effects because they are located in a weaker gravitation field than on the earth, but travel at a considerably higher velocity. This can be corrected by off-setting the clocks before launch, but there is also a periodic drift because of the deviation of the satellite orbit from the circular shape. This deviation gives amplitudes up to about 70 ns at an eccentricity of 0.03, which has to be corrected for as the specification is 1 ns.

The specifications can be met if the corrections in the receiver are modelled as a second-order polynomial of the form

$$\Delta t_{sv} = a_{f0} + a_{f1}(t - t_{0c}) + a_{f2}(t - t_{0c})^2 + \Delta t_r \tag{12.1}$$

and

$$t = t_{sv} - \Delta t_{sv} \tag{12.2}$$

where t = GPS system time, Δt_r is the relativistic correction term (NATO, 1987), t_{sv} is the satellite time at the transmission of the message, t_{0c} is a reference time given by the data block (i.e. the time from the last epoch of the system time, e.g. the time from the last transition from Saturday to Sunday GMT) and a_{f0}, a_{f1} and a_{f2} are coefficients given by the data block I (Table 12.2). (All times are given in seconds.)

Equations (12.1) and (12.2) are coupled by the parameter t, but it turns out that t in Equation (12.1) can be replaced approximately by t_{sv} without essential loss of accuracy.

12.5.3.2 Age of data/group delay

The block also contains the difference between the reference time t_{0c} and the time at the previous updating from the master control centre, so that the user may take the age of data into account (e.g. by weighting them). The normal updating interval of the message is 4 h,

Table 12.2 Data block 1 parameters
Subframe 1 parameters

Parameter	No. of bits**	Scale factor (LSB)	Effective range***	Units
Code on L_2	2	1		dimensionless
Week No.	10	1		week
L_2 P data flag	1	1		dimensionless
Satellite accuracy	4			
Satellite health	6	1		dimensionless
T_{GD}	8*	2^{-31}		s
$IODC$	10			
t_{oc}	16	2^4	604 784	seconds
a_{f2}	8*	2^{-55}		s/s^2
a_{f1}	16*	2^{-43}		s/s
a_{f0}	22*	2^{-31}		s

Notes:

*Parameters so indicated are two's complement, with the sign bit (+ or −) occupying the MSB.

**See Figure 12.6 for complete bit allocation in subframe.

***Unless otherwise indicated in this column, the maximum range attainable with indicated bit allocation and scale factor.

Ionospheric parameters (subframe 4)

Parameter	No. of bits**	Scale factor (LSB)	Effective range***	Units
α_0	8*	2^{-30}		s
α_1	8*	2^{-27}		s/semi-circle
α_2	8*	2^{-24}		s/semi-circle^2
α_3	8*	2^{-24}		s/semi-circle^3
β_0	8*	2^{11}		s
β_1	8*	2^{14}		s/semi-circle
β_2	8*	2^{16}		s/semi-circle^2
β_3	8*	2^{16}		s/semi-circle^3

Notes:

*Parameters so indicated are two's complement, with the sign bit (+ or −) occupying the MSB.

**See Figure 12.7 for complete bit allocation in subframe.

***Unless otherwise indicated in this column, effective range is the maximum range attainable with indicated bit allocation and scale factor.

but in certain situations, e.g. in case of jamming, older data may occur. Block-II satellites can store 14 days of uploading data.

For two-frequency receivers, also, the difference in group delay between the L_1 and L_2 signals up to the satellite antenna is transmitted.

12.5.3.3 Corrections of ionospheric delays

The fundamentals for the description of ionospheric influence and the two-frequency correction method were given in Section 9.1. However, this method necessitates the utilization of L_2. Receivers without this possibility must use a model of the ionosphere in order to reduce the influence, at least to some extent, as the maximum worst-case delay at L_1 may be more than 200 ns.

The ionospheric model developed for one-frequency users of GPS (Klobuchar, 1987) consists of a cosine representation of the daily fluctuation in TEC, where the period and the amplitude of the cosine function depend on the latitude. Studies have shown that an average of 5 ns ('DC term') should be used in addition, and that the maximum of the function occurs at two o'clock p.m. local time.

The coefficients to be transmitted from the satellites are computed from a global, empirical model and updated every 10 days (every 5 days if the sun is particularly active). The model contains a poly-nomial with eight coefficients, denoted α_0, α_1, α_2, α_3 and β_0, β_1, β_2, β_3. (These coefficients are transmitted in subframe 4 (Figure 12.7), but they are presented in this section.) In order to use the model the user has to compute the obliqueness of the signals from the satellite, i.e. the position where the signal path intersects the average iono-sphere altitude. This altitude has been set to 350 km. In addition, the user has to compute his own geomagnetic latitude (as the ionosphere is a function of the geomagnetic field). Accurate instructions about the computations (and with suitable approximations) have been published (Klobuchar, 1987). The final formula is

$$\Delta t = F\left[5 \times 10^{-9} + \sum_{i=0}^{3} \alpha_i \varphi_M^i \left(1 - \frac{x^2}{2} + \frac{x^4}{24}\right)\right] \qquad (12.3)$$

where φ_M is the geomagnetic latitude, E is the elevation angle of the satellite,

$$F = 1 + 16(0.53 - E)^3 \qquad (12.4)$$

$$x = \frac{2 \cdot \pi \cdot (t - 50\,400)}{\sum_{i=0}^{3} \beta_i \varphi_M^i} \qquad (12.5)$$

(All times are in seconds, all angles have been normalized to 180°.)

It appears that this model gives an average reduction of the ionospheric error by 55–60 per cent (Feess and Stephens 1987; Jorgensen, 1989).

12.5.4 Data block 2

12.5.4.1 Introduction

Data block 2 appears in the subframes 2 and 3 (Figure 12.6) and is repeated every 30 s. It is generated by the control segment and contains a parameter representation of the orbit of the satellite and the age of these parameters (Table 12.3). Definitions and explanations of the parameters are given in Tables 12.4 and 12.5.

Table 12.3 Data block 2 (ephemeris) parameters

Parameter	No. of bits**	Scale factor (LSB)	Effective range***	Units
IODE	8			
C_{rs}	16*	2^{-5}		m
Δn	16*	2^{-43}		semi-circles/s
M_0	32*	2^{-31}		semi-circles
C_{uc}	16*	2^{-29}		rad
e	32	2^{-33}	0.03	dimensionless
C_{us}	16*	2^{-29}		rad
$(A)^{1/2}$	32	2^{-19}		$m^{1/2}$
t_{oe}	16	2^4	604 784	s
C_{ic}	16*	2^{-29}		rad
$(OMEGA)_0$	32*	2^{-31}		semi-circles
C_{is}	16*	2^{-29}		rad
i_0	32*	2^{-31}		semi-circles
C_{rc}	16*	2^{-5}		m
ω	32*	2^{-31}		semi-circles/s
OMEGADOT	24*	2^{-43}		semi-circles/s
IDOT	14*	2^{-43}		semi-circles/s

Notes:

*Parameters so indicated are two's complement, with the sign bit (+ or −) occupying the MSB.

**See Figure 12.6 for complete bit allocation in subframe.

***Unless otherwise indicated in this column, the maximum range attainable with indicated bit allocation and scale factor.

Table 12.4 Orbital parameter definitions

M_0	Mean anomaly at reference time
Δn	Mean motion difference from computed value
e	Eccentricity
$(A)^{1/2}$	Square root of the semi-major axis
$(OMEGA)_0$	Longitude of ascending node of orbit plane at weekly epoch
i_0	Inclination angle at reference time
ω	Argument of perigee
$OMEGADOT$	Rate of right ascension
$IDOT$	Rate of inclination angle
C_{uc}	Amplitude of the cosine harmonic correction term to the argument of latitude
C_{us}	Amplitude of the sine harmonic correction term to the argument of latitude
C_{rc}	Amplitude of the cosine harmonic correction term to the orbit radius
C_{rs}	Amplitude of the sine harmonic correction term to the orbit radius
C_{ic}	Amplitude of the cosine harmonic correction term to the angle of inclination
C_{is}	Amplitude of the sine harmonic correction term to the angle of inclination
t_{oe}	Reference time of ephemeris
$IODE$	Issue of data (ephemeris)

The equations given in Table 12.5 (cf. Equations (A7.58)–(A7.64)) provide the satellite's antenna phase centre position in the WGS 84 earth-centered earth-fixed reference frame defined as follows (NATO, 1987):

The origin is the earth's centre of mass (i.e. geometric centre of WGS-84 ellipsoid).

The z-axis is parallel to the direction of the Conventional International Origin (CIO) for polar motion, as defined by the Bureau International de L'Heure (BIH) on the basis of the latitudes adopted for the BIH stations (i.e. rotation axis of WGS-84 ellipsoid).

The x-axis is the intersection of the WGS-84 reference meridian plane and the plane of the mean astronomic equator, the reference meridian being parallel to the zero meridian defined by the Bureau International De L-Heure (BIH) on the basis of the longitudes adopted for the BIH stations.

The y-axis completes a right-handed earth-centred, earth-fixed ortho-gonal coordinate system, measured in the plane of the mean astro-nomic equator 90° east of the x-axis.

Thus, the x-, y- and z-axes are those of the WGS-84 ellipsoid.

The parameters have been selected with reference to the external forces influencing the satellite orbits. Table 12.6 gives an overview of these forces and their influences, and it is obvious that the dominat-ing disturbance is the deviation of the earth from spherical shape.

Second in magnitude are the lunar and solar gravitation forces which vary depending on the position of the satellite. Those are practically constant in short time intervals but can have a considerable influence at minimum distance from the satellite. Together, the forces of Table 12.6 make the satellite follow slightly eccentric elliptic orbits with multiperiodic variations, where the periods are partly the orbital time of the satellite itself (or half of this), and partly the periodic movements of the celestial bodies.

Data block 2 occupies the words 3–10 (parity included) of the subframes 2 and 3. The number of bits, the least significant bit (always the bit which is received last), the magnitude of the figure and the unit are specified in Table 12.3.

12.5.4.2 Parameter representation selection

A number of factors had to be considered at the selection of the parameters describing the GPS satellite orbits. The most important considerations concerned how the selection influenced the situation of the user and the cost of receiver equipment, e.g. the time of the first position computation after start, computation time in the receiver, storage requirements in the receiver, the time validity of the para-meters, ephemeris accuracy, accuracy of the deviation from the nominal ephemeris, rate of degradation (i.e. if there is a smooth degradation of the parameter values or not) and the time needed for transmission of information about other satellites (the almanac). The parameter selection concerned alternatives as time polynomials, har-monic series expansions and Keplerian parameters with perturbations. The latter were selected because they have an obvious physical implication and a smoothly decreasing validity as a function of time. The disadvantage of Keplerian parameters is the computer time in the receiver.

In addition to these parameters, harmonic perturbations are trans-mitted, as shown by Tables 12.3–12.5 (the coefficients C_{uc}, C_{us}, C_{rc},

Table 12.5 Coordinate system elements

$$\mu = 3.986005 \times 10^{14} \ \frac{\text{meters}^3}{\text{sec}^2}$$

WGS 84 VALUE OF THE EARTH'S UNIVERSAL GRAVITATIONAL PARAMETER

$$\dot{\Omega}_e = 7.2921151467 \times 10^{-5} \ \frac{\text{rad}}{\text{sec}}$$

WGS 84 VALUE OF THE EARTH'S ROTATION RATE

$$A = (\sqrt{A})^2$$

SEMI-MAJOR AXIS

$$n_0 = \sqrt{\frac{\mu}{A^3}}$$

COMPUTED MEAN MOTION

$$t_k = t - t^*_{oe}$$
$$n = n_0 + \Delta n$$
$$M_k = M_0 + n t_k$$
$$M_k = E_k - e \sin E_k$$

TIME FROM EPHEMERIS REFERENCE EPOCH
CORRECTED MEAN MOTION
MEAN ANOMALY
KEPLER'S EQUATION FOR ECCENTRIC ANOMALY (MAY BE SOLVED BY ITERATION)

$$v_k = \tan^{-1}\left\{\frac{\sin v_k}{\cos v_k}\right\}$$

TRUE ANOMALY

$$= \tan^{-1}\left\{\frac{\sqrt{1 - e^2}\sin E_k/(1 - e\cos E_k)}{(\cos E_k - e)/(1 - e\cos E_k)}\right\}$$

$$E_k = \cos^{-1}\left\{\frac{e + \cos v_k}{1 + e\cos v_k}\right\}$$

ECCENTRIC ANOMALY

Table 12.5 *Cont.*

$\phi_k = v_k + \omega$	ARGUMENT OF LATITUDE
$\delta u_k = C_{us} \sin 2\phi_k + C_{uc} \cos 2\phi_k$	Argument of Latitude Correction ⎫
$\delta r_k = C_{rc} \cos 2\phi_k + C_{rs} \sin 2\phi_k$	Radius Correction ⎬ SECOND HARMONIC
$\delta i_k = C_{ic} \cos 2\phi_k + C_{is} \sin 2\phi_k$	Correction to Inclination ⎭ PERTURBATIONS
$u_k = \phi_k + \delta u_k$	CORRECTED ARGUMENT OF LATITUDE
$r_k = A(1 - e\cos E_k) + \delta r_k$	CORRECTED RADIUS
$i_k = i_0 + \delta i_k(IDOT)t_k$	CORRECTED INCLINATION
$x'_k = r_k \cos u_k$	
$y'_k = r_k \sin u_k$. ⎫	POSITIONS IN ORBITAL PLANE
$\Omega_k = \Omega_0 + (\dot{\Omega} - \dot{\Omega}_e) t_k - \dot{\Omega}_e t_{oe}$	CORRECTED LONGITUDE OF ASCENDING NODE
$x_k = x'_k \cos \Omega_k - y'_k \cos i_k \sin \Omega_k$ ⎫	
$y_k = x'_k \sin \Omega_k + y'_k \cos i_k \cos \Omega_k$ ⎬	EARTH FIXED COORDINATES
$z_k = y'_k \sin i_k$ ⎭	

*t is GPS system time at time of transmission, i.e. GPS time corrected for transit time (range/speed of light). Furthermore, t_k shall be the actual total time difference between the time t and the epoch time t_{oe}, and must account for beginning or end of week crossovers. That is, if t_k is greater than 302.400 seconds, subtract 604.800 seconds from t_k. If t_k is less than 302.400 seconds, add 604.800 seconds to t_k.

Table 12.6 Forces distorting the GPS satellite orbits and their approximate influence

Source	Maximum acceleration of the satellite (m/s²)	Maximum change of satellite position during 1 hour (m).
Earth gravitation (G_e)	0.565	–
The second harmonic of G_e	5.3×10^{-5}	300
Lunar gravity	5.5×10^{-6}	40
Solar gravity	3×10^{-6}	20
Solar radiation pressure	10^{-7}	0.6
Gravitation anomalies	10^{-8}	0.06
All other forces	10^{-8}	0.06

Source: van Dierendonck *et al.*, 1978.

C_{rs}, C_{ic}, C_{is}). These are the second harmonics of G_e (Tables 12.7) caused by the deviation of the earth from spherical shape. All the parameter values are valid for 1.5 h.

The equations given in Table 12.5 give the position of the phase centre of the satellite antenna in Cartesian coordinates with the z-axis along the rotation axis of the earth and the x-axis in the equatorial plane, in the direction of 0-meridian (Equations (A7.58)–(A7.62)).

12.5.5 Data block 3

This data block (Table 12.6) comes in subframe 5 but is not repeated every 30 s as are the earlier blocks. There are twenty-five data groups in block 3 which follow by turns in subframe 5 such that it takes a total of 12.5 min to transmit everything.

Data block 3 is generated by the control segment and contains the almanac, i.e. the coarse ephemerides of the other satellites (up to twenty-four), and the clock correction parameters, satellite identification (specifying which of the thirty-two codes is used by that specific satellite) and health (Figure 12.6). This is to enable the receiver to acquire the other satellite signals directly after locking to the first one. The ephemerides and the clock parameters of block 3 are truncated versions of the block 1 and 2 data messages of that specific satellite. This is shown by a comparison (the scale factor and the number of bits) between Tables 12.7 and 12.2–3. The parameters, which are not shown in Table 12.7, are kept to zero in computations according to Table 12.5. Further, the clock drift is assumed to be linear (the coefficient a_{f2} is zero).

Table 12.7 Data block 3 (almanac parameters)

Parameter	No. of bits**	Scale factor (LSB)	Effective range***	Units
e	16	2^{-21}		dimensionless
t_{oa}	8	2^{12}	602 112	s
δi****	16*	2^{-19}		semi-circles
$\dot{\Omega}$	16*	2^{-38}		semi-circles/s
$(A)^{1/2}$	24	2^{-11}		m$^{1/2}$
Ω_0	24*	2^{-23}		semi-circles
ω	24*	2^{-23}		semi-circles
M_0	24*	2^{-23}		semi-circles
a_{f0}	11*	2^{-20}		s
a_{f1}	11*	2^{-38}		s/s

Notes:

*Parameters so indicated are two's complement, with the sign bit (+ or −) occupying the MSB.

**See Figure 12.6 for complete bit allocation in subframe.

***Unless otherwise indicated in this column, the maximum range attainable with indicated bit allocation and scale factor.

****Relative to $i_0 = 0.30$ semi-circles.

The reference time of the almanac, t_{0a}, is the multiple of 4096 seconds nearest below 3.5 days after the beginning of the almanac transmission. GPS system time (which covers only 1 week) never deviates more than 3.5 days from t_{0a}. The almanac is renewed every 6 days.

The accuracies of the satellite positions decrease with aging of the data and are about 3200 m after 1 week and about 24000 m after 5 weeks. But the latter distance is only about 820 bit-lengths of the P-code (82 of the C/A-code) and is thus a good means of acquisition of the signals from the other satellites.

12.6 Receivers

12.6.1 Introduction

With the exception of situations where the altitude is known (or for special uses as, for example, interferometry and time transfer), a GPS receiver has to be able to receive and track as well as measure the times of arrival of signals from at least four satellites. The general tasks of a receiver with regard to each spread-spectrum signal have been described in Section 11.2.5. In addition the navigation message must be demodulated and utilized, and the final (and for many uses the only) visible result is the presentation of the position and other navigational data. The increased computational power of all navigational receivers makes it possible to present a mass of data which were not available to the user before; this, of course, also applies to GPS. Usually, the computational unit of a receiver is separated into two main parts: the receiver processor and the navigation processor. The first part controls the receiver function, whereas the other part executes the navigational computations.

12.6.2 Receiver types

A characterization of receivers can be performed in different ways. Use of the number of channels of the receiver and the applications is common. The latter usually states whether the receiver should stand high accelerations or not (high or low dynamics), which is mainly determined by the properties of the phase-locked loops and their auxiliary functions. The number of channels reveals the solution of the problem to track several satellite signals. Airborne equipment has particularly strict requirements for dynamic properties, especially in missiles and fighter aircraft. Civil aircraft have less strict requirements in this respect, and man-pack and geodetic receivers, as well as shipborne receivers, have still lower dynamics requirements.

The number of channels varies between:

- multichannel receivers (usually two, five or more parallel channels);
- sequential receivers;
- multiplex receivers.

In all these receivers the RF part is independent of the receiver type. If several channels share the same signal, a separation is made at IF or baseband. A multichannel receiver has several parallel, identical receiver channels so that it can track several satellites simultaneously without interruption. Such a solution is more expensive and is mainly used in receivers with high dynamics requirements (Tachita *et al.*, 1989) and in geodetic-type receivers. A fifth channel in a four-channel receiver (or a second in a one-channel receiver) may be used to enable the receiver to acquire a new satellite without interrupting the tracking of another in order not to disturb the position computation, or to allow simultaneous measurements on the other carrier.

Time sharing can be used instead of a multichannel receiver, mainly to reduce costs but also to reduce weight and volume. There are two ways of utilizing this method: sequential and multiplex reception.

A typical sequential receiver tracks each satellite for about 1 s. The necessary time for full position determination is thus 4–5 s. Such a receiver can never perform as well as a multichannel receiver, but in a static situation the difference is minimal. In a dynamic situation, on the other hand, it gives an effective signal-to-noise-ratio 2–8 dB lower than a multichannel receiver (Maher, 1984). The difference depends on the receiver design but is due to the fact that the sequential receiver has to acquire a new satellite each second, and acquisition needs higher S/N than tracking. In order to collect data a sequential receiver has to track a certain satellite for 6 s to receive a whole sub-frame (Section 12.5.1) (Tachita *et al.*, 1989). This does not mean reduced S/N but has to be carried out in situations with low dynamics to avoid problems in tracking the other satellites. The transients of the transition to a new satellite should be short; for this reason a bandwidth of the code loop should be less than 50 Hz. In high dynamic situations, a sequential receiver may need aiding (e.g. by an inertial system) to be able to accomodate the accelerations. The reason is that at the beginning of a new dwell on a satellite, a sequential receiver carries out a small search and acquisition of the code and carrier phase errors which have built up since the last dwell a few seconds ago, and makes code and carrier measurements after the settling of the loops.

A multiplex receiver goes through a whole cycle, i.e. four or five satellites, in the course of one bit-length of the data message (20 ms). Thus it is about 200 times faster than sequential receivers. It samples the signals only once in each cycle and uses the sampled values for updating the software of the tracking loops. Because of the high sampling rate, the multiplex receiver can both read data and search for new satellites while outputting navigational data. A rule-of-thumb is that the sampling frequency in feedback control systems (like the code and carrier loops of the receiver) has to be at least three times the bandwidth. A third-order phase-locked loop with a bandwidth of 15 Hz enables the receiver to track satellite signals, even if it is carried by a fighter aircraft (Maher, 1984). A second-order code-tracking loop not aided by a carrier loop would require a bandwidth of 3–6 Hz, but with rate-aiding from a carrier loop the bandwidth can be less than 1 Hz, thus reducing ranging noise.

Some of the most important properties of multiplex receivers are as follows:

- Both the code- and the carrier-tracking loops track the signals of each satellite continuously through the software.

- The state variables of each satellite signal are stored in a program loop to be used in the next tracking interval (this applies to position, velocity and acceleration).

- At the beginning of a dwell the calculated code phase and frequency are loaded into the tracking loop so that the desired code state is realized in a fraction of one C/A code length (<400 μs (Maher, 1984)); the same thing is also done with the phase and frequency of the carrier.

- The distribution of the track intervals of the different satellites is made by the receiver with reference to receiver time so that bit transitions in the data signal never occur within one track interval. (There are solutions which do not require this.) This time control is continuously adjusted so that the order between the satellites can be changed without losing lock in the carrier loop.

- As all the satellite signals are received through one channel, the multiplex receiver is less sensitive to receiver channel variations than is a multichannel receiver.

Multiplex reception can also be realized with the P-code, and even with SA (selective availability, Section 12.9) running. In relation to a multichannel receiver a multiplex receiver requires about 25 per cent more processor capacity (Maher, 1984), and with the accuracy requirement in addition it is necessary to use digital receivers (see below). In comparison to a multichannel, or a sequential receiver a multiplex receiver has lower S/N because it tracks each satellite for only part of the time (four satellites gives -6 dB, five satellites -7 dB, etc.).

In multichannel, as well as multiplex receivers, carrier tracking is used to support the code tracking. The frequency gradient of the carrier-tracking loop makes it possible to compute the corresponding gradient in the code-tracking loop so that the control signal of the VCO in the latter loop can be computed. This also enables the receiver designer to use a reduced bandwidth in the code-tracking loop, from, for example, 3–6 Hz in a second-order loop without such aid, to less than 1 Hz with such aid. A further reduction of bandwidth can be achieved if integrated Doppler information (as in TRANSIT) is utilized.

12.6.3 Receiver description

The generalities of spread spectrum and the fundamentals of spread-spectrum receivers were given in Chapter 11. This section deals with some more details and gives examples from some commercial receivers.

An example of analogue receiver design, i.e. a receiver where all signal processing is analogue until the processing of recovered data is shown in Figure 12.8 (Daly and Allison, 1986).

A general tendency since the beginning of GPS receiver development has been an increasing use of digital solutions. This can be illustrated as in Figure 12.9 where the arrows show possible levels of digitization.

Digitization of the signals is simpler if it is done after, rather than before, the correlation process. This reduces bandwidth requirements considerably, with a corresponding increase in S/N. The danger of interference and intermodulation between different satellite signals is also eliminated. The signal can either be digitized with a bandwidth of 15–20 kHz without consideration of the Doppler shift which is eliminated by subsequent signal processing, or the Doppler shift can

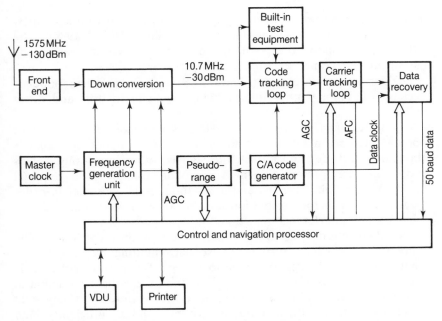

Figure 12.8 Example of analogue receiver block diagram

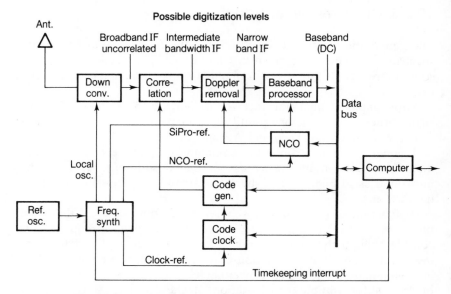

Figure 12.9 General GPS receiver concept (Holte, 1989)

be eliminated before digitization, by means of mixing with a synthesized LO frequency tracking the Doppler shift (Holte, 1989). The latter method allows the possibility of further S/N improvement by bandwidth reduction.

Digitization at baseband (Figure 12.8) is becoming less popular, because of, among other things, high sensitivity to drift and offset voltages.

Digitization at IF before the correlator seems likely to be the most popular method in future developments. The C/A-code spectrum is about 2 MHz wide, and all satellite signals, other interfering signals and noise are present. A high sampling frequency is required (20–50 MHz (Holte, 1989)), and a 12–16-bit A/D converter in order to achieve satisfactory resolution and linearity for code-phase discrimination and interference rejection.

The digitization itself can be performed in different ways. The simplest and cheapest way is to use hard limiting by means of a computer (1-bit digitization) (Mattos, 1988; Beier and Wolf, 1987; Trimble, 1988). Disadvantages of this method are a slight reduction in S/N (about 2 dB) and increased sensitivity to interference. Using A/D converters with a sufficient number of bits and a sufficiently high sampling frequency is more expensive but gives a better result.

Recently, the transputer has been proposed for use in GPS receivers (Mattos, 1989). The transputer is a very high-speed general-purpose processor with on-chip serial communications that can transfer up to 1.8 Mbytes/s on each of eight links. Operations such as ADD or XOR needed for GPS signal processing take 50 or 100 ns. The transputer can handle multiple jobs at the same time, which is a definite advantage in a multichannel receiver.

Digitization of the signal and fast microprocessors with large memories enable the receiver functions to deviate considerably from corresponding analogue receiver functions. It is possible, for example, to store the received signal digitally in a time interval of one or several code lengths (C/A) followed by a correlation of the stored signal with locally generated versions of all possible satellite signals in parallel (Mattos, 1988). This gives a very short acquisition time, a few milliseconds, but requires a fast processor and a large memory.

The carrier-tracking loop and parts of the code-tracking loop can be implemented by means of FFT after correlation as such a process can inform the correlator of the reception of a satellite signal when a certain part of the signal spectrum exceeds a preset threshold; and simultaneously the FFT gives the phase of this spectral component. In order to achieve a smaller bandwidth and consequently a better signal-to-noise ratio the FFT window can be controlled by the output signal in a simple way. The detected phase is utilized in the data demodulation, and the level of the mentioned spectral component is used as input data to the digital code-tracking loop to ensure centering on the correlation peak.

Figure 12.10 shows the function block diagram of a C/A-code multiplex receiver where digitization takes place after the correlator. A more detailed schematic of the same receiver, mainly used for differential GPS (Section 12.7) is shown in Figure 12.11. One particular feature to be noted is the use of the bi-phase method for code tracking. In this method two hardware channels are used for tracking, one for the carrier and the other for the code (Holte, 1989). Thus, channel differences do not cause errors. The code-phase discriminator is realized by modulo-2 addition of the code and the 50 per-cent-duty-cycle clock. Then the first and the second half of each code chip contribute to the correlation result with opposite signs. At full synchronism the two contributions balance each other, otherwise the sign and magnitude of the unbalance denote the sign and the magnitude of the code-phase error.

A commercial receiver with additional features for different applications is shown schematically in Figures 12.12 and 12.13 (Trimble, 1988).

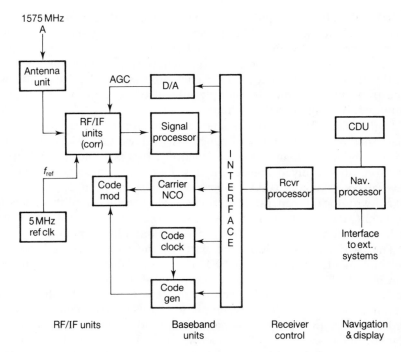

Figure 12.10 Multiplex receiver block diagram (courtesy Kongsberg Navigation AS)

The receiver front end in Figure 12.13 consists of an antenna, a bandpass filter, an amplifier and a mixer. The antenna choice (Section 12.7) depends on the application, but in Trimble (1988) a quadrifilar helix is suggested. It is followed by a band-limited low-noise, two-stage GaAs-FET amplifier with typically 35 dB gain, and a balanced image-reject mixer outputting a signal at nominally $4f_0$ where $f_0 = 1.023$ MHz. The filter following the mixer is a six-pole bandpass filter at $4f_0$ with 3 dB bandwidth of 2 MHz. After that the signal is amplified by 40–50 dB and hard-limited.

For the code- and carrier-tracking loops, a phase-controlled C/A-code signal synthesizer and a frequency-controlled Doppler-shifted carrier-signal synthesizer are employed. The code synthesizer feeds the early-late counter through modulo-2 additions of the received signal. The code selected for the synthesizer is read from an addressed code-storage memory. The carrier synthesizer generates two digital signals in phase quadrature and with $\frac{1}{4}$ Hz resolution. Both loops are locked through the microcomputer.

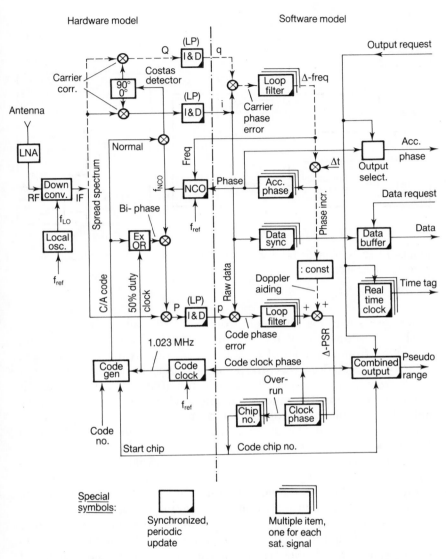

Figure 12.11 Fundamentals of the KV/PS GPS D-set (courtesy Kongsberg Navigation AS)

The 12-bit up/down counters in the I and Q branches count the number of agreements minus the number of disagreements between the synthesized and the received codes. The early–late counter is activated (up or down) when one, and only one, of the synthesized

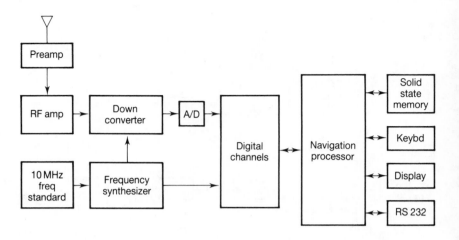

Figure 12.12 Schematic block diagram of multichannel receiver (courtesy Trimble Navigation)

codes agrees with the bit stream from the hard limiter.

The time-of-arrival of the satellite signal is derived by the microcomputer from the control signal to the code synthesizer, and the velocity information is correspondingly derived from the carrier synthesizer control signal. The navigational message bit stream is obtained from the I counter, and this is also used for inverting, or not, the sign-bit signal of the Q counter. The filter after the Q counter indicated in the diagram thus represents the loop filter of a Costas loop for carrier tracking.

The microcomputer also compensates for variations in the downconverted satellite signal level to maintain the carrier loop gain constant. (The I counter output rather reflects the S/N, but as the noise level is fairly constant, it also reflects the signal level.)

A substantial part of the digital portion of the receiver in Figure 12.13 is integrated in a gate-array-type device and in a custom-design LSI-type device (Trimble, 1988). Increasing use of custom design LSI is typical of modern GPS receivers.

Finally, it should be added that the above basic functions of a receiver are supplemented in one way or the other, depending on the application. Often some form of carrier aiding is used for reduction of code-loop noise (Section 12.6.2). In receivers for differential use (Section 12.7), pseudo-range corrections received from a reference site are included in the position computations. A corresponding use

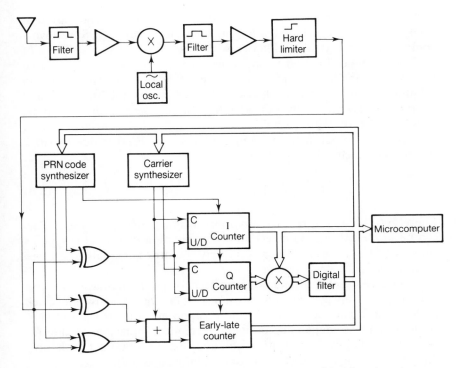

Figure 12.13 Function block diagram of L_1 multichannel receiver (courtesy Trimble Navigation)

of phase reference data is necessary for geodetic-type high-accuracy receivers (Section 12.8). The latter receivers also contain means for detecting, and possibly correcting, carrier-phase slips.

12.6.4 Clock and frequency stability

As the determination of the slant range to each satellite is based on TOA measurements of the signals of the satellite with reference to the user clock, it is obvious that this clock plays a significant role for the measurement result. This was also indicated by the inclusion of the clock error (with regard to satellite time, in the derivation of position errors in Section 9.2.2). The clock error influences the position error through the geometrical factors (Equations (9.46), (9.47) and (9.51)).

A clock is in reality a stable frequency reference, and different reference oscillators have different stabilities. Most GPS receivers contain quartz crystal oscillators with short-term (1 s) stabilities of $10^{-9}-10^{-11}$ and long-term stabilities (days, weeks) two or three orders of magnitude worse. More expensive receivers may contain rubidium or even cesium clocks with short-term stabilities as good quartz clocks but with considerably better long-term stabilities, $\approx 10^{-12}$ and $\approx 10^{-13}$, respectively.

Every slant range measurement contains a time error due to clock inaccuracy which can be written

$$\Delta t = \int_0^t \frac{\Delta f(t)}{f}\, \mathrm{d}t \tag{12.6}$$

where $\Delta f(t)/f$ is the relative clock error as a function of time. This can be modelled, e.g. in a second-order equation of the same type as Equation (12.1), but however accurately the drift is modelled, a noise term remains.

A good clock may attenuate the effects of poor satellite geometry, but there is always an unambiguous relationship between clock performance, satellite geometry and position accuracy (Copps, 1984). A clock can also, depending on its performance, replace a satellite for some time, i.e. a 3-D position solution can be obtained with only three satellites available (Knable and Kalafus, 1984–5).

The clock influences more than the position accuracy. As shown in the preceding section, various intermediate frequencies are obtained by mixing with LO signals derived from a frequency reference, usually by heterodyne techniques. Noise or frequency drift on these signals will degrade the performance of the receiver. For example, excessive frequency drift complicates the search process and may even prevent acquisition. This is one of the reasons why receivers need a certain warm-up time before acquisition can take place. Most receivers contain temperature stabilization circuitry for the reference frequency crystals.

Another form of frequency drift, but much slower, is ageing. This changes the specific frequency, which was calibrated at manufacture, as a function of time and use. The result may be increased initial acquisition time, and even phase-lock difficulties.

12.6.5 Antennae

The satellite signal is right-hand polarized, and the satellite antenna is an array antenna consisting of single helices as elements. One of the most common receiver antennae is also a (quadrifilar) helix with four microstrip lines with half-turns on a cylindrical substrate (Kilgus, 1969; Laube, 1987) (Figure 12.14). This half-turn, half-wave antenna

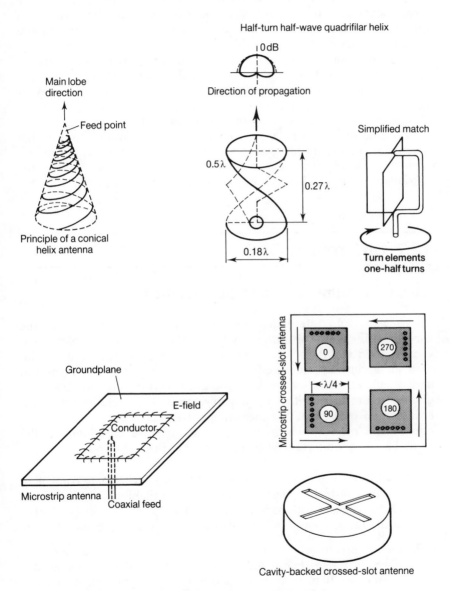

Figure 12.14 Antennae for GPS receivers

has a relatively small bandwidth, about 20 MHz at L_1 frequency. A larger bandwidth, covering both L_1 and L_2, can be obtained with another type of quadrifilar helix, a logarithmic helix on a conical substrate (Figure 12.14). The signal output is at the peak of the cone which is about 7 cm high and has a base diameter of about 10 cm. A quarter-wave monopole (4.7 cm high), can be used as well, but has 3 dB polarization loss.

Another common antenna is a quadratic microstrip patch antenna asymmetrically fed to realize the circular polarization. This is a relatively cheap antenna, but its polarization properties are not very good at low elevation angles (gives elliptical polarization). At 5° elevation it is 2 dB worse than the quadrifilar helix.

For aircraft and similar applications it is desirable to use flat antennae, preferably integrated into the fuselage. Compound microstrip antennae, cavity-backed crossed dipoles or other low-profile antennae (Figure 12.14), can be utilized (Smith *et al.*, 1989).

12.7 Differential GPS

12.7.1 Presumptions and possibilities

The reason for differential use of GPS is the same as in the corresponding case of differential OMEGA and LORAN-C, and translocation in TRANSIT. A reference receiver is located at a point with an accurately measured position. By comparison of measured and calculated results at this position, corrections are produced for distribution to users within the coverage area. This technique works if an essential portion of the measurement errors is caused by factors outside the receiver which at the same time are correlated in the area. The most important of these errors are the following:

1. Deliberately generated errors in the satellite signals controlled by the system operator (Selective Availability, SA, Section 12.9). These errors have an assumed standard deviation of 40 m in the range measurements. PS users do not suffer from this.

2. Ionospherical delays of the signals. These can be as large as 100 m in range (Feess and Stephens 1987) and the σ values are around 20–30 m in daytime and 3–6 m at night, but with great variations as functions of time and position. Two-frequency

receivers can practically eliminate this error as it is inversely proportional to the frequency squared (Section 3.5).

3. Tropospheric time delays. At low elevations, even these can be as large as 30 m, but they are relatively regular and can be calculated by means of empirical models (Section 3.4).

4. Ephemeris errors, i.e. differences between the actual satellite position and the position given by the navigation message. Normally these are only a few metres, but in some cases they may be 20–30 m.

5. Satellite clock errors, i.e. differences between the actual satellite time and the time given by the navigation message. These errors can be as large as 5–10 ns, i.e. a few metres in range.

Some of errors 1–5 above can be eliminated by differential methods; this applies to direct clock errors as well as to clock errors generated as a part of SA. (However, it has to be remembered that such clock errors are changed at relatively short time intervals, so that corrections have to be recent.) The influence of other errors is reduced, and the reduction decreases with increasing distance between the reference station and the user. Ionospheric and tropospheric errors have varying correlation distances, in time as well as in position. The satellite elevation also plays a role here. Ephemeris errors have different influence at different positions because the geometry depends on the user position. Thus, radial, along-track and across-track errors have different influence. Maximum influence occurs when the user and reference receivers are in the same plane as the correct and erroneous positions of the satellite. Figure 12.15 shows that maximum ephemeris errors are proportional to the distance between the reference station and the user.

There are principally two types of corrections which may be transferred from the reference stations, namely corrections of measured distances and corrections of measured positions. The latter type requires less transmission capacity but induces errors if the user does not utilize the same satellites as the reference station. For this reason the best results are achieved when the reference station transmits range corrections to all satellites above a certain elevation angle, such that the user can optimize his own satellite selection and utilize only those corrections which are valid for the selected satellites.

Correction data transmission rates are determined by the variation rates of the errors which are to be corrected. The most uncertain part in that connection is played by SA.

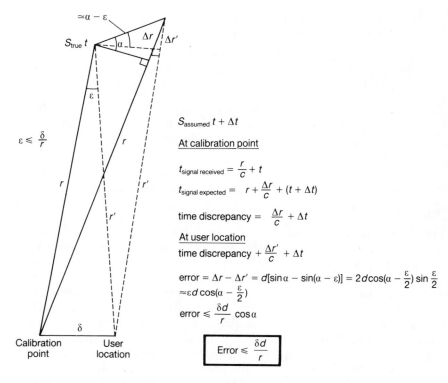

At calibration point

$$t_{\text{signal received}} = \frac{r}{c} + t$$

$$t_{\text{signal expected}} = r + \frac{\Delta r}{c} + (t + \Delta t)$$

time discrepancy $= \frac{\Delta r}{c} + \Delta t$

At user location

time discrepancy $+ \frac{\Delta r'}{c} + \Delta t$

error $= \Delta r - \Delta r' = d[\sin\alpha - \sin(\alpha - \varepsilon)] = 2d\cos(\alpha - \frac{\varepsilon}{2})\sin\frac{\varepsilon}{2}$
$\approx \varepsilon d\cos(\alpha - \frac{\varepsilon}{2})$

error $\leqslant \dfrac{\delta d}{r}\cos\alpha$

$$\boxed{\text{Error} \leqslant \frac{\delta d}{r}}$$

Figure 12.15 Maximum influence by ephemeris errors in differential GPS

12.7.2 Work up to now

Many countries have worked on differential GPS for several years, the most comprehensive work being carried out by the RTCM Special Committee 104 (Radio Technical Commission for Maritime Services), which has worked out proposals for standardization of reference station messages (Kalafus *et al.*, 1986). These proposals are based on the same principles as the navigation message from the satellites, mainly to enable a user to utilize so-called pseudolites (Klein and Parkinson, 1984–5). These are fixed transmitters on the ground which transmit in the same way as ordinary satellites in orbit and which are to be utilized in special situations, e.g. to improve system coverage and reliability in limited areas.

SC104 proposals contain several message types which can be used in different situations so that the message length is kept at a minimum. There are, however, certain requirements to be complied with. There are sixteen different message types envisaged in the format. Like the satellite message it is divided into subframes, and the wordlength is thirty bits. Time is also given in the shape of Z-counts (Section 12.5.2), but with 0.6 s as a scale factor instead of 6 s. The transmission speed and the data rate are determined by SA to a large extent, and the requirements would be considerably less strict without this perturbation (Figure 12.16). The bit rate is 50 b/s, mainly to fit into the satellite message. The transmission channel itself has not been decided, but many options are given, among others the modulation of signals from maritime beacons (Enge *et al.*, 1987). Another option is the use of LORAN-C signals (Section 4.5.9) (Forssell, 1985; Hoogenraad, 1988).

Considerable work has also been carried out in Norway to develop differential GPS (Hervig, 1988). Tests have shown (Forssell, 1984) (Figures 12.17–12.19), that performance improvements compared with ordinary C/A-code receivers are considerable. The measurements were carried out in southern Norway with the reference point on a central building roof at The Norwegian Institute of Technology, Trondheim. Differential errors well below 10 m indicated that a differential system with two or three reference stations could cover

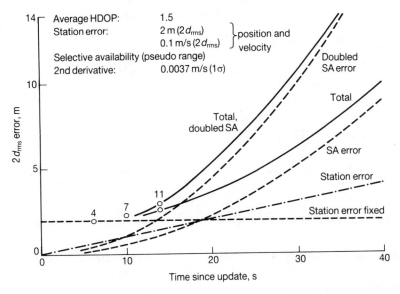

Figure 12.16 Errors as a function of time in differential GPS due to SA (Kalafus, 1989)

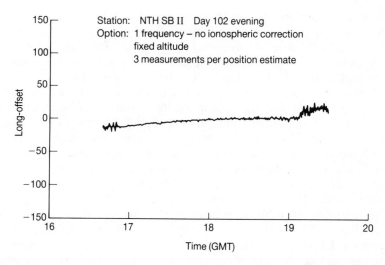

Figure 12.17 Measurement results from Trondheim with GPS/CA-code in two dimensions and with 6 s of averaging. Longitudinal errors in meters

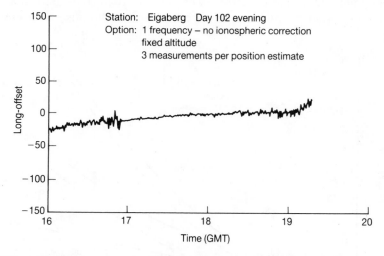

Figure 12.18 The same as Figure 12.17 but from Eigaberg at Stavanger

the whole of Norway with 10 m accuracy. Accuracies can be further improved by utilization of integrated Doppler corrections, (down to the 2–4 m range).

Because of these measurements a system called DiffStar, with reference transmitters on islands in the north of Norway and one

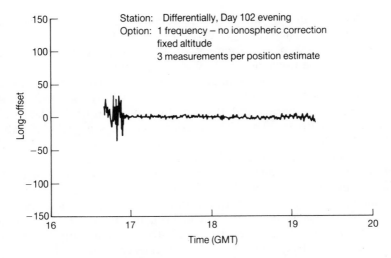

Figure 12.19 Differences between measurements in Figures 12.16 and 12.17

outside Bergen, has been developed (Hervig, 1988). Tests have been run since 1987, and the system is now operational. On one northern island the old Consol transmitter is utilized to transmit corrections (at the carrier frequency 332.5 kHz), and in the Bergen area a similar transmitter is used. The signal format deviates a little from the proposal by SC-104, mainly because pseudolites are not assumed. Range corrections are updated every 10 s. The system is commercial and used particularly in oil and gas activities off the coast.

12.8 Interferometric use

Geodetic precise positioning requires accuracies in the sub-metre area, which cannot be satisfied by differential GPS. The GPS applications discussed above are based on the measurement of times-of-arrival of a certain point of the code, i.e. really code-phase measurement. As was shown earlier (Section 4.2), the accuracy (in radians) achievable is inversely proportional to the square root of the signal-to-noise ratio (S/N). As a certain phase accuracy corresponds to a range accuracy which is proportional to the wavelength, a given S/N implies that the range accuracy increases proportionally with frequency. This fact has led to the utilization of carrier-phase measurements in the case of strict requirements for resolution and range accuracy. The typical value of $S/N = 30$ dB in the carrier-tracking loop gives a phase accuracy of $1.8°$ which at L_1 (wavelength 19 cm), means a range accuracy of about 1 mm.

C/A-code measurements imply an ambiguity problem because the code period is 1 ms. This is solved either by knowledge of the user's own position within about 150 km or by utilization of the data signal whose bit length is 20 ms. The carrier phase has a period of about 0.6 ns, but the corresponding ambiguity is more difficult to solve. The phase is always measured with reference to the receiver located at a reference point to which the user wishes to measure his range. The phase difference between the measurement and reference points can be many periods, i.e. the range difference to a satellite is many wavelengths. Users having access to both the P- and C/A-code, i.e. both the L_1 and L_2 frequencies, can solve the ambiguity by utilizing the difference frequency $L_1 - L_2$ and the code frequency, which, however, requires a fixed, calibrated link between the measurement and reference points, which rarely exists. The normal procedure is therefore to utilize the satellite's own movement, i.e. to measure for a sufficient period so that satellite position changes can be utilized, and/or to make more range difference measurements, possibly to different satellites, followed by taking the difference between the results (see below) (Forssell, 1988; Remondi, 1986).

The methods can be summarized in the following way.

1. Single difference (Figure 12.20). This is a measurement of the phase difference between the measurement and reference points at one point in time. The method eliminates satellite-dependent clock errors and reduces the influence of ephemeris errors

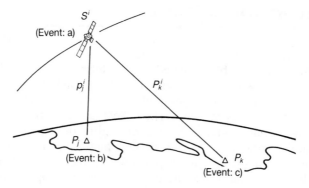

Figure 12.20 Single difference measurement

(Figure 12.15). Range ambiguity is not eliminated, nor are differences in clock errors between the two receivers. Ionospheric errors are eliminated if the baseline between the receivers is short. In the case of many such measurements (several hundreds), the range ambiguities, the coefficients of a second-order polynomial describing the difference between the drift parameters of the receiver clocks and the difference between the position coordinates of the receivers can be estimated by means of the method of least squares.

2. Receiver-time difference measurement, also called Doppler difference (Figure 12.21). This implies two measurements of type 1 at two different points of time. The method eliminates satellite errors and time-independent errors. Even the ambiguity is eliminated if the period between the two difference measurements is not too large. The disadvantage of the method is small measurement values and thus poor measurement accuracy.

3. Double difference (Figure 12.22) The method implies measurements of type 1 but with the use of two different satellites. The difference between the measurement results is utilized, and in the ideal case both satellite errors and receiver errors then vanish and only time-independent errors remain. As in the case of type 1, coordinate differences, integer ambiguities and clock drift coefficients can be estimated, using the method of least squares. The number of parameters to be estimated is, however, considerably less than for type 1, and thus the requirements for computation capacity as well. Therefore this method is the most common among those mentioned.

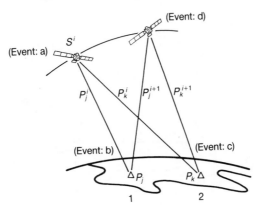

Figure 12.21 Doppler difference measurement

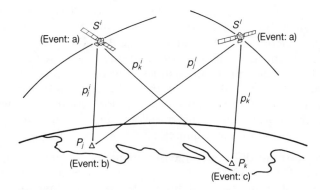

Figure 12.22 Double difference measurement

4. Triple difference (Figure 12.23). The method implies the same as type 3, but at two different points of time, with the difference between the results being utilized. In the ideal case this method eliminates time-independent errors and integer ambiguity, but it suffers from the same weaknesses as type 2, i.e. poor measurement accuracy.

All the methods described above are very demanding with regard to the execution of the measurements and imply considerable post-processing of measurement data. (Real-time cm-accuracy measurements are, however, being investigated, and some progress is being made.) Measurements give better results in static conditions, but they can also be executed with

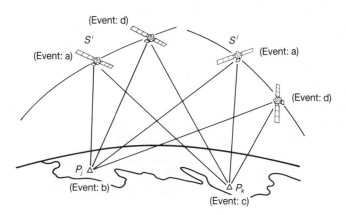

Figure 12.23 Triple difference measurement

moving receivers. The more information about the measurement situation and parameters (satellite orbits and clocks, receiver clocks, ionospheric and tropospheric conditions, etc.), the better the results. Costs of such measurements are often considerable. Residual errors are mainly caused by receiver noise and multipath interference (signal reflections).

12.9 Accuracy

It has been made clear in the preceding sections that the accuracy of the system depends on several factors which the user himself can influence, e.g. measurement time, receiver design, post-processing of measurement data and selection of point of measurement and point of time (for static measurements). There are, however, also conditions influencing the accuracy which the user himself cannot influence, e.g. access to L_2 and the P-code, satellite constellation and integrity and reliability of the system. (By integrity we mean the ability of the system to report on parameters not being within specifications, and by reliability we mean the probability of the occurrence of such situations (Kalafus, 1989).)

The accuracy of GPS is divided into two classes: SPS and PPS, meaning Standard and Precise Positioning Service respectively. According to system specifications, SPS in the operational system will give a horizontal accuracy of 100 m in 95 per cent of all cases, and 140 m vertically. The corresponding figures for PPS are 18 m and 27 m. Whether a user can utilize the possibilities of PPS or not depends on, firstly, if he wishes to (receivers are more expensive), and, secondly, if he meets the PPS user requirements from the system management. The situation is that knowledge of the P-code does not have any automatic implications, because this will be encrypted in the operational system (the Y-code, Section 11.3) such that users are dependent on a decryption key (NATO, 1987). Rules and instructions have already been published for those who are interested in using PPS (Larkin, 1988).

SPS is based on the C/A-code. However, tests have shown that obtainable accuracy with this code is about 36 m (95 per cent) horizontally. Therefore the system management (i.e. in reality the US DOD), has decided that errors, in particular the ephemeris and clock data of the navigation message, should be added deliberately so that 100 m (95 per cent) will be the result. This is the disreputable

Selective Availability (SA). The reason is that so-called non-authorized users will not be able to utilize the system with disadvantages to the United States. Details about this accuracy reduction are difficult to find, but Figure 12.16 gives some indication. The error distribution of the 5 per cent of all measurements which do not give results within 100 m is also unknown. Figure 12.24 shows accuracy curves for SPS and PPS and what can be achieved with the C/A-code alone without SA.

GPS can also be used for transmission of accurate time (better than 100 ns). That accuracy will also be reduced by SA, so it has been proposed that a few satellites (one or two in each orbit) be used for this purpose without being disturbed by SA (Green *et al.*, 1989). (It is enough in this case that the two receivers to be synchronized have one good satellite in view at the same time.) However, the discussion about this is still going on, and at the time of writing it seems unlikely that any block-II satellite will be exempted from SA.

The accuracy figures given above are according to specifications. Measurements have shown that the real performance (without SA) is somewhat better. Since the 1970s a number of tables and curves have been published concerning theoretical and measured accuracy data under various circumstances and the influence by different error sources (e.g. Parkinson and Gilbert, 1983; Blackwell, 1985; Mard *et al*, 1989; Hagle, 1988). Tables 12.8 and 12.9 show error budgets for

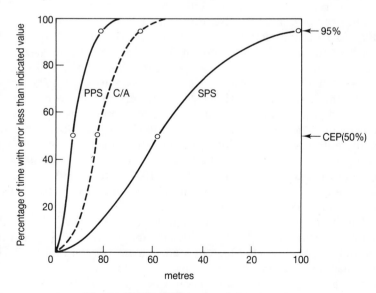

Figure 12.24 GPS horizontal accuracies

Table 12.8 Static error budget for GPS (without SA)

Error source	Expected range measurement error (m rms)	
	P-code	C/A-code
Satellite clock errors	1–3	1–3
Ephemeris errors	2.5–7	2.5–7
Ionospheric delay (residual after correction)	0.4–2	2–15
Tropospheric delay (residual after correction)	0.4–2	0.4–2
Noise and quantizing in the receiver	0.1–0.3	1–2
Differences between receiver channels	0–0.1	0–0.2
Multipath propagation	1–2	2–4
Resulting range error in the receiver	3–8	4–18
Resulting position error horizontally (HDOP = 1.5) vertically (VDOP = 2.5)	4.5–12 7.5–20	6–27 10–45

Table 12.9 Static error budget for differential GPS, 90 km between the receivers

Error source	Expected range measurement error (m rms)	
	P-code	C/A-code
Residual satellite clock error	0	0
Residual ephemeris error	0–0.1	0–0.1
Residual ionospheric and tropospheric errors	0–0.5	0.1–1.5
Differences between receiver channels	0–0.1	0–0.1
Multipath propagation	1.5–3	2–5
Receiver noise	0.1–0.4	1–3
Resulting range errors in the receiver	1.5–3	2–6
Resulting position error horizontally (HDOP = 1.5) vertically (VDOP = 2.5)	2–5 4–8	3–9 5–15

usual static measurements and for differential measurements.

A corresponding error budget for interferometric measurements depends on several additional factors, concerning measurement method, time and processing. But, as was mentioned in Section 12.8, it is possible to obtain relative accuracies down to the mm-range, if the effort is good enough. It is usual to express such relative accuracy as the ratio between the error and the length of the baseline, in ppm (parts per million). Today's technology can give 1 ppm, but there are growing requirements from geodetic users for 0.1 ppm. Today's ephemeris data given by the navigation message are, however, too inaccurate to permit this, so special tracking stations have to be established which distribute precise coordinates of the satellites from accurately known positions to possible users (Forssell, 1988; Engen, 1988). Such high-precision users are normally not dependent on real time results, which makes it possible to utilize methods independent of SA.

12.10 Integration with other navigation systems

Even with twenty-four satellite positions, constellations with poor satellite geometry (e.g. high values of GDOP and PDOP) may occur in some situations. In high-dynamic situations the receiver loops may have difficulty in accommodating the accelerations. A turning aircraft may also lose lock of one or more satellites. Shadowing by houses, bridges, mountains, etc., may occur rather frequently in land navigation (Held and Kricke, 1985). In all these situations, position errors from GPS can be reduced to an acceptable level by using other navigation systems as aids. In general, combination of any navigation system with other systems through Kalman filtering (Schmidt *et al.*, 1989) will always reduce the navigational errors of the former.

12.10.1 GPS and inertial navigation systems

Many navigation systems have been suggested for integration with GPS. Most of the investigations of such integration have dealt with inertial systems (INS). The reason for this is that GPS and INS have complementary features. If the GPS receiver loses lock because the antenna is shadowed or because of high acceleration, an INS is capable of providing accurate aiding data on short-term vehicle movements, so that the GPS receiver can either keep track of the satellite signal through a dynamic manoeuver or re-acquire the signal very quickly. INS aiding can also be utilized for maintaining relatively low tracking bandwidths, thus enabling the receiver to withstand high noise levels, e.g. jamming. The dominating drawback of INS is the low-frequency drift, velocity errors of the order of 1–2 m/s, and position errors of about 2 km/hour, but when GPS and INS are integrated, this drift can be compensated for, and the INS can be realigned and recalibrated by means of GPS.

The Kalman filter algorithms (Appendix 9) for such integration usually incorporate a fairly high number of variables or states, e.g. position, velocity and acceleration in three coordinates, clock bias and drift, accelerometer errors, gyro drift parameters, etc., often twelve to twenty parameters (Hansen, 1989). The algorithms frequently require substantial execution times (Cox, 1978).

It should be noted that most GPS receivers already have a Kalman filter to process the raw pseudo range and range-rate measurement data obtained from the code- and carrier-tracking loops respectively. Filtered position and velocity values are usually output at one or a few second intervals to aid the tracking logic. A Kalman filter is based upon modelling the vehicle kinematics, often a constant velocity model disturbed by constant acceleration between measurement updates (Napier, 1989). The constant acceleration is often modelled as an unbiased random variable with Gaussian distribution whose variance can be changed by the user in some receivers, thereby accounting for different dynamic applications. The existence of such filters in GPS receivers has implications for integration with INS.

The optimum solution is full integration of GPS and INS, using one centralized Kalman filter. This filter includes the effects of satellite geometry and models of GPS clock errors and INS errors. It can also accept pseudo ranges from less than four satellites and detect and reject bad pseudo ranges. The filter designed to accomplish these tasks is a complex one but can provide full navigational capability also if either GPS or INS fails. However, it requires a special-purpose GPS receiver. There are many variations with regard to detailed solutions, e.g. to overcome stability problems and to match the system to the constraints and priorities of the application. It is important to note that integration with INS can also improve the GPS performance in differential and interferometric modes. In the latter case it can help remove the influence of cycle slips (Eissfeller and Spietz, 1989).

12.10.2 GPS and LORAN-C

Combining GPS and LORAN-C will improve the performance of either system alone (Braisted et. al, 1986), but because the properties of LORAN differ from those of INS, the mechanization, filter solutions and results will be different.

LORAN signal propagation velocities vary as a function of surface conductivity (Section 4.5). Most receivers attempt to correct for these errors by models, but these models are not very accurate due to geographical, diurnal and seasonal variations. Using GPS as a means of calibration, this situation can be improved considerably.

GPS can also aid LORAN in case of problems in locking to the correct zero crossing and in solving ambiguity problems (i.e. when two hyperbolas intersect at more than one point).

On the other hand, LORAN can aid GPS in situations of satellite shadowing or other outages, by adding one or more LOPs. After having been accurately calibrated at a known position, by differential GPS or other means, a LORAN-C receiver drifts very slowly (1–3 m/hour) because of the slow variation of propagation conditions (Sæther, 1988). Thus, LORAN can be used for improving GPS navigation results for several hours after such calibration.

12.10.3 Other combinations

As indicated above, integration of GPS with any other navigation system would improve the navigation performance. However, in many cases the improvement would be minor, and it must always be evaluated against costs and complexity.

On this background integration with OMEGA is not very cost efficient. Both systems are global, but OMEGA has a considerably poorer accuracy. Such integration would help only when no other aids are available (e.g. in oceanic navigation). The OMEGA outputs would then be kept calibrated by means of good GPS solutions, so that fairly good OMEGA measurements (comparable to differential OMEGA) could be utilized in periods of bad satellite geometry or GPS failures.

Another combination which has been proposed is GPS and a barometric altimeter (Stein, 1985). The advantage of this combination is that the altimeter gives information about the vertical position component, and GPS VDOP is usually larger than HDOP because of the constellation. Thus, baroaltimeter aiding is geometrically more favourable than 'horizontal' aids such as LORAN or OMEGA. This can also give a mathematically better solution to the problem of selecting the constellation with the smallest GDOP. The accuracy of a barometric altimeter is comparable to the pseudo-range measurement error in GPS, being on an average a factor two or so larger (Stein, 1985).

The combinations of GPS and other systems described above are not the only possibilities. Others involve, for example, VOR, DME, direction finders and TRANSIT. However, because of its complementary performance INS is one of the most interesting GPS partners, but also the most expensive. It should finally be mentioned that integration of GPS and GLONASS receivers is a most interesting possibility for the future.

12.11 Perspectives

During the many years which have passed since the initial design of GPS, an enormous amount of knowledge about satellite navigation and positioning in general has been acquired, and not only about GPS itself. This is also reflected in other system projects, and is perhaps the system's most important contribution so far (see next chapter). From a user point of view, geodetic applications and improvements of earlier results in this field have had the greatest implications, but in future such applications, though very important, will represent only a small part of possible applications.

Uncertainty about (above all) SA, but also administrative inaccessibility and lack of integrity, are likely to delay, and may prevent, the proliferation of system use which would otherwise be possible.

Because of the Space Shuttle disaster in January 1986 GPS was delayed more than two years in relation to previous plans. Since then, problems with Delta rockets have meant still more delays to the system, and full operability is not envisaged until the middle of the 1990s.

This disaster also more or less eliminated the shuttle as the main GPS satellite launcher. Delta-II rockets are now used, and the launching scheme indicates five new (Block-II) satellites per year. After the system has become operational, the average launch rate is expected to be once every third month, for replacement purposes.

13 GLONASS

13.1 Introduction

GLONASS (Global Navigation Satellite System) is the future Soviet satellite navigation system, expected to be operational in the mid-1990s (Anodina, 1988a). It has been under development since the 1970s but was first officially announced in February 1982 when the Soviet Union informed the International Telecommunication Union of its intention to establish a new global satellite navigation system (Anodina, 1982). The first satellite in this system was launched on 12 October 1982 and subsequent launches have usually taken place twice each year (Dale and Daly, 1987). In 1989–90 there was a remarkable halt in launching for one year, probably due to recent satellite failures. Short satellite lifespan has been common, usually not more than two years and often much less.

GLONASS satellites are launched three at a time by Proton launchers from the Tyuratam Space Centre in Central Asia (Daly, 1988). Usually, official information about launches and system development have been scarce, maybe only a few lines in *Pravda*. The GLONASS satellites are usually given numbers in the Cosmos series. Until 1988 most of the information about GLONASS was published by a research group at the University of Leeds (e.g. Dale and Daly, 1987; Daly, 1988; Daly *et al.*, 1988; Dale and Daly, 1988), but then the Soviet Union started to promote the system for international use, and data were released officially (Anodina, 1988a). There is now an agreement between the United States and the Soviet Union about efforts in the direction of common use of GPS and GLONASS for

322

civil aircraft navigation. Work is also underway to design integrated receivers for combined use of both systems. And ESA and INMAR-SAT are examining the possibilities of monitoring both GPS and GLONASS by means of special payloads in geostationary satellites (Joyce *et al.*, 1989; Rosetti, 1989). After noting all this, however, it should not be forgotten that GLONASS has about the same military features as GPS, and civil use of the GLONASS 'P-code' (see below) has never been indicated by the system providers. On the other hand, Soviet officials have repeatedly stated that anything similar to Selective Availability will never be implemented on GLONASS.

13.2 Satellite orbits

The GLONASS satellites use circular orbits 19,100 km above the earth, and the orbital time is 11 h 15 min 44 sec. (Anodina, 1988b). The inclination is 64.8°. In full operation the system will consist of a twenty-four-satellite constellation, including three spares, with seven or eight satellites in each of three orbit planes 120° apart. The initial implementation period will lead to a constellation of ten to twelve satellites in two orbit planes 240° apart, making a daily period of 10–18 h (Anodina, 1988b).

The orbit produces a ground-track repeat every seventeen orbits (Daly, 1988). (This should be compared to the GPS repeat every two orbits, before the orbit change (Section 12.2).) The orbital time implies that the diurnal time displacement with regard to a solar day is 4.07 min, i.e. the seventeen orbits last 8 days less 32.56 min. Consequently, a satellite performs every day $\frac{17}{8}$ revolutions + $\frac{1}{8}$ revolution (= 45°). Thus, over a ground-track repeat interval of 8 days, all satellites in the same plane appear in turn at the same position at intervals of 1 day minus 4.07 min. This also implies that the resonance phenomenon provoking the (Section 12.1) change of the GPS orbit does not occur.

The angular phase difference between satellites in adjacent orbital planes is (±)30°. This, together with the in-plane phase difference of 45°, implies that in one 8 day period, all satellite footprints pass through a given position. In other words: if a ground narrow-beam antenna is pointed at one satellite at a certain point in time, all the other satellites will pass through the beam within 8 days.

13.3 The navigation signals

Like GPS, GLONASS uses two carrier frequencies modulated by two spread-spectrum codes and a data sequence. The satellites transmit right-hand circularly polarized signals by means of array antennae with 10–13 dB gain. Contrary to the GPS case, each satellite has its own specific 'L_1' and 'L_2' frequency. The L_1 frequencies are (Anodina, 1988b):

$$f_{L1} = f_0 + i \cdot \Delta f_{L1} \quad i = 1, 2, \ldots, 24 \tag{13.1}$$

where

$$f_0 = 1602 \text{ MHz}$$

and

$$\Delta f_{L1} = 0.5625 \text{ MHz}$$

The relationship between L_1 and L_2 frequencies is (Daly, 1988)

$$f_{L1}/f_{L2} = 9/7 \tag{13.2}$$

The satellite antenna beamwidths are about 30°, with a slight rise of antenna gain at the beam edges. The user link budget for the L_1 signal is as shown in Table 13.1. (Contrary to GPS, P-code and C/A-code powers are equal.)

Table 13.1 Link budget for the Glonass L_1 user

Satellite EIRP {	boresight	25 dBW
	at ± 15°	27 dBW
Space loss		182–184 dB
Atmospheric and tropospheric loss		1–2 dB
Receiver antenna gain		−2–0 dB
Received power		−156–161 dBW
Noise power density		−204 dBW/Hz
Receiver noise factor		4 dB
Receiver signal-to-noise density ratio (P_1/N_0)		39–44 dBHz
Information data bit rate		50 bits/s
Signal-to-noise ratio (E_b/N_0)		22–27 dB

13.4 The codes

The GLONASS codes are very similar to those of GPS, but because of the frequency division all satellites can use the same codes. The code rates are 0.511 Mbits/s and 5.11 Mbits/s for the 'C/A'- and 'P'-code respectively.

Soviet officials have never indicated that the P-code (we will stick to GPS terminology) will be available to external users, so only the C/A-code (PRSA in GLONASS terminology) is treated in the following. This code is binary-phase modulated onto the carrier as the pseudo random part of the modulo-2 sum of the code and the data signal (Anodina, 1988b). The code is generated in the common way by a feedback shift register, as shown in Figure 13.1. Because of the shift register length, the length of the code sequence is $2^9 - 1 = 511$ bits. Consequently, the code repetition rate is 1 kHz.

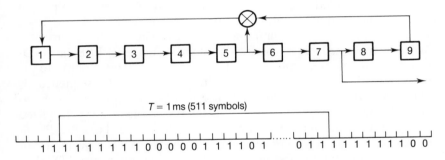

Figure 13.1 The GLONASS C/A-code generator

The P-code seems to be a fairly simple truncated PN code (Lennon, 1989). The received signals indicate that it is generated by a 25-bit shift register with feedback from the 3rd and 25th outputs (Figure 13.2). A maximum-length code from such a generator has a length of $2^{25} - 1 = 33554431$ bits. With the clock rate of 5.11 Mbits/s, the code period would be 6.5664 seconds, but it is short-cycled once a second (Lennon, 1989), i.e. the code vector is set to all ones on the second epoch. Hence the Glonass P-code repeats every second.

Speculation about the simplicity of this code compared to the GPS P-code (Section 11.3) have led to the conjecture that encryption by frequency hopping in the Glonass channelized frequency scheme is planned.

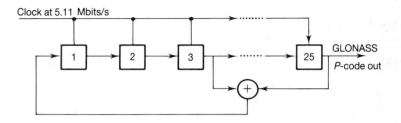

Figure 13.2 GLONASS P-code generator

13.5 The navigation message

Navigation data are bi-phase modulated onto the carrier at 50 bits/s. The sequence is composed of two binary sequences and has a total length of 2 s, called one line (Anodina, 1988b). The first 1.7 s part of a line is the modulo-2 sum of the data message (DD) and a 010 . . . sequence, the latter alternating at 100 bits/s. The last 0.3 s part of a line is a truncated binary pseudo random sequence (PRSB in GLONASS terminology), running at 100 bits/s (i.e. 30 bits long), to be used for time synchronization. The actual PRSB sequence is 111110001101110101000010010110. The first DD bit (i.e. at the beginning of one line) is always zero. All bit transitions are synchronized (as in GPS), within 2 ms for all satellites. The last PRSB bit represents a time mark in even integer seconds from the beginning of the day, Moscow time. (Sometimes, another time reference, UT(SU), i.e. Soviet Union Universal Time, has been used.)

The DD structure is formed by superframes of length 2.5 min, each superframe consisting of five frames of 30 s, and each frame consisting of fifteen lines (subframes). The last 8 bits of each line are Hamming parity bits, and each line starts with a 32-bit preamble. The DD words are always transmitted with the most significant bit first. The first four lines in every frame contain the ephemeris, health and clock-correction parameters; the fifth line contains a UTC correction parameter and the almanac day number. Lines 6–15 contain almanac data, distributed on five frames. Thus, the complete navigation message takes 2.5 min. The details of the data format can be found in Anodina 1988a, b.

There are some significant differences between the navigation messages of GPS and GLONASS. Both give precise orbital information about the transmitting satellite's own status and position, and lower-precision information about other satellite positions. Whereas

GPS satellites give their positions in Kepler parameters (Section 12.5), GLONASS positions are given in earth-centred, earth-fixed x-, y-, z-coordinates. In addition to coordinates, GLONASS also gives velocity and (lower-accuracy) acceleration. This method facilitates receiver computations but implies a shorter time validity of the information, forcing the user to interpolate or extrapolate to find position and velocity at other than reference times. GLONASS satellite positions are given every half-hour, and ground-control ephemeris update is made every day. GPS almanacs are valid for several days, but GLONASS almanacs are usually updated every day. However, there is a greater similarity between almanac than ephemeris data as GLONASS almanacs are given as Kepler parameters also.

There are other differences also between GPS and GLONASS messages. Most of these concern satellite status, but it should be noted that the GLONASS format allows a few more parameters to be included. No ionospheric correction parameters are given.

13.6 Receivers

In principle, GPS and GLONASS receiver complexity should be comparable. The main difference stems from the use of code and frequency-division multiplex respectively. Research is ongoing to develop combined receivers, both military and civil (Daly and Lennon, 1989; Dale and Daly, 1988).

13.7 Accuracy

The Soviet Union's official position is that GLONASS is designed for civil maritime and aircraft use, giving position accuracies of 100 m (95 per cent) and time accuracy of 1 μs (Anodina 1988b). This seems to be a modest figure, and trials indicate that the potential is at least as good as GPS (Daly et al., 1988; Moskvin and Sorochinsky, 1989). Soviet officials have repeatedly said that there will be nothing like SA (Section 12.9) in GLONASS. Combined receivers to take advantage of both systems look, superficially at least, to be attractive. However, improved accuracies, reliability and integrity are unlikely to be attained unless many issues are satisfactorily solved. These include final GPS orbit, level of SA, time differences, geodetic references and susceptibility to interference. (It should be kept in mind that GLONASS in particular is still in an experimental phase.)

14 Other satellite navigation systems

14.1 Introduction

It might not seem very logical to establish or run other satellite systems as GPS and GLONASS would have the possibilities of covering most needs. However, as we have seen, both are mainly military systems, with civil use being only a secondary task. This might be called the institutional aspect and has been discussed for many years and in many fora (e.g. Forssell, 1986; McGann, 1987; Johannessen, 1989), among these ICAO (the FANS committee) and IMO (the Sub-Committee on Safety of Navigation). Cost aspects are of course an essential part of that question, bearing in mind that GPS implementation costs are of the order of $10 billion and running costs about 10 per cent of that per year. Civil use of both systems is for free, so far, but it should be remembered that GPS satellites do have cryptography equipment on board (after Congress demand) in order to introduce user charges for C/A-code users also.

Moreover, 100 m accuracy far from covers all accuracy requirements. Differential and interferometric use is an expression of the requirements of more and more users and would have come in any case, but SA worsens the situation.

In addition, system integrity from a civil point of view is not adequate as GPS and GLONASS performance is not monitored for civil users. This seems to exclude these systems from the primary navigational aids recommended by ICAO, even though there are proposals to use other communication channels for monitoring, e.g. by INMARSAT satellites (Joyce *et al.*, 1989) or other geostationary satellites (Kalafus, 1989; Rosetti, 1989). It is, however, doubtful whether satisfactory monitoring would be enough as signal availability is not guaranteed under all circumstances.

Finally, GPS and GLONASS cannot be utilized for localization (i.e. position reporting) without additional means, which has generated some development work, as will be seen below. This issue might be considered as part of the question concerning combination or common utilization of communication and navigation channels.

Below, some other satellite systems for navigation and positioning are superficially described. There are other systems which could be included as well, but those mentioned are representative of what is going on.

14.2 TSIKADA

This is the Soviet equivalent of TRANSIT. Like TRANSIT it will be phased out when its successor (i.e. GLONASS) is operational, i.e. towards the end of the 1990s. It has existed since the mid-1960s (Perry and Wood, 1976) but has been used very little by non-Soviet users. The system is described here for the sake of completeness and has nothing to do with the aspects discussed in Section 14.1.

The TSIKADA principles of operation resemble those of TRANSIT. This applies to the use of two carrier frequencies, 150 and 400 MHz, polar orbits with one revolution in about 105 min, and Doppler techniques for user position determination. Whereas 400 MHz is the main frequency for TRANSIT, it is 150 MHz for TSIKADA.

TSIKADA also has military and civil tasks. The system generally has twelve satellites in operation (this may vary somewhat): eight for military and four for civil purposes (Daly, 1984).

Although the data transmission rate is 50 bits/s, i.e. near the TRANSIT rate, the modulation techniques differ. TSIKADA uses low-frequency tones for the primary modulation followed by amplitude modulation onto the carrier. The tone frequencies are 3, 5 and 7 kHz. The first two tones are used differentially to give a zero or a one in the binary information, and a 7 kHz synchronizing pulse marks the start of a new second (Daly, 1984). The information encoded uses principles similar to GLONASS (Section 13.4), i.e. satellite position and velocity in x-, y-, z-coordinates, Moscow time and Keplerian parameters of the other satellites.

14.3 STARFIX

This is a privately owned and operated positioning system, operational since 1986 (Ott and Blanchard, 1988). It has been created mainly to serve the oil industry in the Gulf of Mexico but covers the continental United States as well. The principle implies pseudo range measurements to three geostationary satellites with known positions. Signals are transmitted from ground up-link stations through the satellites and down to the users. A network of ground stations controlled by a Master Site tracks the satellites and makes the system operate in a pseudo differential mode. Up-link and down-link frequencies are about 6 GHz and 4 GHz respectively. The passive user equipment determines satellite ranges by means of PRN-codes of 16,383-bit lengths transmitted at 2.4576 Mbits/s. The data rate is 150 baud, and the code repeats 150 times per second. The stated position accuracy is 5 m ($2d_{rms}$).

14.4 GEOSTAR/LOCSTAR

This is a system based on similar principles as STARFIX, the main differences being the frequencies and the fact that GEOSTAR/LOCSTAR is a two-way ranging system with two-way information transfer capability. GEOSTAR is created and operated by the US company of the same designation (Rothblatt and van Till, 1986), whereas its European equivalent, LOCSTAR (Hernandez, 1986), is promoted by a consortium headed by the French space research organization Centre National d'Etudes Spatiales (CNES). The aim is to develop a system each for coverage of the mid-latitude parts of North America and Europe, the Middle East, Africa respectively.

The principle of position determination implies that ranging signals are transmitted back and forth between the system centre (in Washington, DC, and Toulouse, respectively) and the user via two or three geostationary satellites (only one satellite is used for transmission from the centre) (Figure 14.1). For 3-D position determination, either transmission via three satellites or via two satellites plus height determination by other means is used. As there are four signal paths which are not allowed to interfere, four frequency bands are required. In 1987 the mobile World Administrative Radio Conference

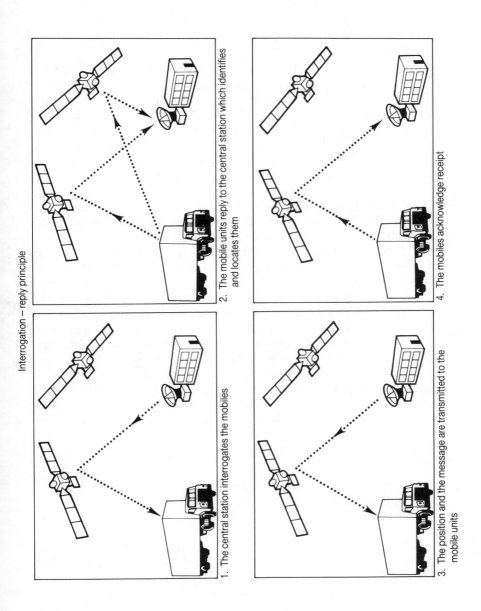

Figure 14.1 The GEOSTAR/LOCSTAR principles

(WARC) allocated the following bands to be used for Radio Determination Satellite Service (RDSS): mobile up-link 1610–1626.5 MHz, mobile down-link 2483.5–2500.0 MHz, and satellite control station link 5150–5216 MHz, with power limited to avoid interfering with other users.

This system also uses PRN codes for range measurement with bit rates of about 8 Mbits/s and a code length of $2^{17} - 1$ bits and Gold codes. The outbound information rate is 62.5 kbits/s and the inbound rate 15.625 kbits/s (Snively and Osborne, 1986).

Because of the two-way transmission, the system has limited capacity. It has been announced that the system will be able to manage 5700 256-bit inbound transmissions per second for an overall system capacity of 1.4 Mbits/s (Rothblatt and van Till, 1986). This figure includes radio-positioning and radio-navigation requests, as well as ancillary messaging. The corresponding number of users is over 2 million per hour. Each user would be able to communicate with the centre and with any other user (via the centre).

The position accuracy is stated to be about 7 m (Rothblatt and van Till, 1986). The horizontal accuracy of 7 m depends on the user knowing his own altitude; if not, a few hundred metres is achievable. Because of the geometry, the equatorial areas obtain lower accuracies, and the polar areas are not covered by geostationary satellites.

Every user paying a few hundred dollars for hand-held equipment and some tens of dollars per month would receive one update per hour, obtaining extended service for more money as an option. The number of users at any moment can be a few thousand.

Because of satellite problems the operational status of Geostar has been delayed. Using one satellite only, the system started to relay LORAN-C position information to interested users in June 1988 (Foley, 1988). Dedicated satellites for both GEOSTAR and LOC-STAR are scheduled for 1992.

14.5 NAVSAT

NAVSAT is a global, civil satellite navigation system planned by the European Space Agency (ESA) (see Section 14.1), and is meant to be a system open to everybody, with different accuracy classes according to the user's requirements and his willingness to buy more or less expensive equipment (Rosetti, 1986). In addition, some extra functions are to be included, e.g. monitoring of GPS and GLONASS (Joyce et al., 1989).

NAVSAT will be based on eighteen satellites, six in the geostationary orbit with 60° spacing (GEO) and twelve in high-elliptical orbits (HEO) (Rosetti, 1986). The latter orbits are proposed to be of the Tundra type with 63.45° inclination, 24-hour orbital period, and the apogee and perigee altitudes 53,632 and 17,861 km respectively. There will be three orbital planes with four satellites and two orbits each. One orbit will have its perigee in the southern hemisphere to cover the areas north of the equator, and the other orbit will have its perigee in the northern hemisphere to cover the areas south of the equator (Figure 14.2). The advantage of this constellation is mainly that it is easy to build up successively to obtain 3-D coverage in selected areas during the building-up period with a small number of satellites.

The choice of carrier frequency has been made principally to facilitate receiver design for reception of GLONASS and GPS monitor signals, in addition to the system's own navigation signals. The carrier frequency therefore fits into the GLONASS scheme (Equation (13.1)) with the 'L_1' frequency close to that of GPS (Rosetti, 1989). GPS and GLONASS look-alike signals are proposed to be modulated in quadrature onto the carrier, with monitor information of the respective system at suitable positions in the data message (Rosetti, 1989). A C-band frequency, modulated in the same way, for ionospheric corrections and high-precision applications (Baltzersen *et al.*, 1988), will be added.

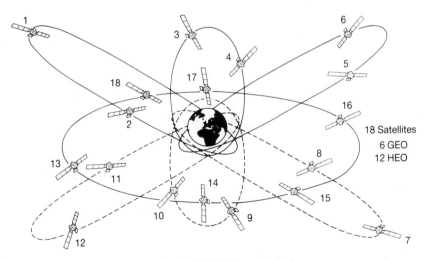

Figure 14.2 NAVSAT constellation

Thus, the modulation signal will consist of a PN-code and a 50-bits/s data message in analogy with GPS and GLONASS. Because modifications of the signal format are still possible at the moment of writing, details of the format are left out here.

Details of the space and control segments and suitable host satellites are yet to be determined.

Appendix 1
Datum transformation

As indicated in Section 1.2 above, different datums are generally characterized by different reference ellipsoids centred on different points. With the assumption that the rotational axis is the same (this assumption is not always correct), the ellipsoids differ by five parameters, namely the lengths of the axes and the coordinates of the centres. This appendix shows in detail the mathematics needed for the transformation of position coordinates between two such datums.

An arbitrary plane containing the rotational axis of an ellipsoid (the z-axis) gives the intersection ellipse

$$\frac{w^2}{a^2} + \frac{z^2}{a^2(1 - e^2)} = 1 \tag{A1.1}$$

where e is the eccentricity and $w^2 = x^2 + y^2$. The geodetic latitude Φ (Figure 1.9) is the angle between the perpendicular to the ellipse at the point in question and the w-axis. The tangent of the ellipse at the same point is found by differentiation of the equation of the ellipse, i.e.

$$\frac{\mathrm{d}z}{\mathrm{d}w} = -\frac{(1 - e^2)w}{z} \tag{A1.2}$$

Thus

$$\tan \Phi = -\left(\frac{\mathrm{d}z}{\mathrm{d}w}\right)^{-1} = \frac{z}{w(1 - e^2)} \tag{A1.3}$$

Substitution of Equation (A1.3) in Equation (A1.1) gives

$$w = \frac{a}{\sqrt{1 + (1 - e^2)\tan^2 \Phi}} = \frac{a\cos\Phi}{\sqrt{1 - e^2 \sin^2 \Phi}} \qquad (A1.4)$$

and

$$z = \frac{a(1 - e^2)\sin\Phi}{\sqrt{1 - e^2 \sin^2 \Phi}} \qquad (A1.5)$$

Assuming that a position in Datum 1 is given by the coordinates Φ_1, Λ_1 and h_1 (where h_1 is the height above the ellipsoid) gives

$$w_1 = w + h_1 \cos\Phi_1 \qquad (A1.6)$$

$$z_1 = z + h_1 \sin\Phi_1 \qquad (A1.7)$$

where w and z are given by Equations (A1.4) and (A1.5) with parameters e_1 and a_1 (Figure A1.1). In a three-dimensional Cartesian coordinate system with the same origin, the coordinates of P in Datum 1 are

$$x_1 = w_1 \cos\Lambda_1 = \left(\frac{a_1}{\sqrt{1 - e_1^2 \sin^2 \Phi_1}} + h_1\right)\cos\Phi_1 \cos\Lambda_1 \qquad (A1.8)$$

$$y_1 = w_1 \sin\Lambda_1 = \left(\frac{a_1}{\sqrt{1 - e_1^2 \sin^2 \Phi_1}} + h_1\right)\cos\Phi_1 \sin\Lambda_1 \qquad (A1.9)$$

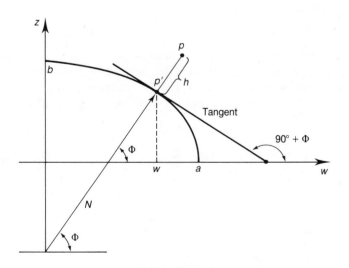

Figure A1.1 Plane section of the ellipse

$$z_1 = \left(\frac{a_1(1 - e_1)^2}{\sqrt{1 - e_1^2 \sin^2 \Phi_1}} + h_1 \right) \sin \Phi_1 \tag{A1.10}$$

Datum 2 differs from Datum 1 by a displacement of the origin of Δx, Δy and Δz along the coordinate axes, respectively, and by the values of the eccentricity and semi-axis of the ellipse. Consequently, in this second system,

$$\left. \begin{aligned} x_2 &= x_1 + \Delta x \\ y_2 &= y_1 + \Delta y \\ z_2 &= z_1 + \Delta z \end{aligned} \right\} \tag{A1.11}$$

$$w_2 = \sqrt{x_2^2 + y_2^2} \tag{A1.12}$$

and

$$\Lambda_2 = \arctan^{-1}\left(\frac{y_2}{x_2} \right) \tag{A1.13}$$

All equations in Datum 2 are of the same form as their equivalents in Datum 1, so according to Equations (A1.8)–(A1.10)

$$w_2 = \left(\frac{a_2}{\sqrt{1 - e_2^2 \sin^2 \Phi_2}} + h_2 \right) \cos \Phi_2 \tag{A1.14}$$

$$z_2 = \left(\frac{a_2(1 - e_2^2)}{\sqrt{1 - e_2^2 \sin^2 \Phi_2}} + h_2 \right) \sin \Phi_2 \tag{A1.15}$$

Equations (A1.14)–(A1.15) give

$$\frac{w_2}{\cos \Phi_2} - \frac{z_2}{\sin \Phi_2} = \frac{a_2 e_2^2}{\sqrt{1 - e_2^2 \sin^2 \Phi_2}} \tag{A1.16}$$

which can be rewritten as

$$\Phi_2 = \arccos\left(\frac{w_2}{\dfrac{z_2}{\sin \Phi_2} + \dfrac{a_2 e_2^2}{\sqrt{1 - e_2^2 \sin^2 \Phi_2}}} \right) \tag{A1.17}$$

As w_2 and z_2 are known quantities (from Equations (A1.11)–(A1.12)), Φ_2 can be computed numerically by iteration of Equation (A1.16). A suitable initial value is $\Phi_{21} = \Phi_1$. After Φ_2 has been computed with sufficient accuracy, h_2 can be found by Equation (A1.14) or (A1.15).

Appendix 2
Formulas from spherical trigonometry

In spherical trigonometry, which deals with angles and distances on spherical surfaces, all quantities are regarded as angular. The physical distances are normalized to the radius of the sphere and consequently expressed by the aspect angles seen from the centre of the sphere. By definition, all arcs (sides) of a spherical triangle can be associated with planes passing through the centre of the sphere and are thus sections of great circles (Figure A2.1).

(Proofs of the following theorems are not given here, but they can be found in Næss, 1952, for instance.)

A2.1 Theorems concerning right-angled spherical triangles

Assuming that the angle at C in Figure A2.1 is a right angle, the side c is denoted hypotenuse and the other sides catheti. The following formulas are valid:

$$\cos c = \cos a \cdot \cos b \qquad\qquad (A2.1)$$

$$\tan a = \tan c \cdot \cos B \qquad\qquad (A2.2a)$$

$$\tan b = \tan c \cdot \cos A \qquad\qquad (A2.2b)$$

$$\tan a = \sin b \cdot \tan A \qquad\qquad (A2.3a)$$

$$\tan b = \sin a \cdot \tan B \qquad\qquad (A2.3b)$$

$$\sin a = \sin c \cdot \sin A \qquad\qquad (A2.4a)$$

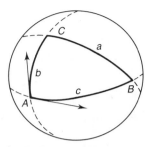

Figure A2.1 Spherical triangle. Capital letters denote angles on the surface, small letters, centre aspect angles, (i.e. superficial distances normalized to the radius of the sphere)

$$\sin b = \sin c \cdot \sin B \qquad\qquad\text{(A2.4b)}$$

$$\cos a = \frac{\cos A}{\sin B} \qquad\qquad\text{(A2.5a)}$$

$$\cos b = \frac{\cos B}{\sin A} \qquad\qquad\text{(A2.5b)}$$

$$\cos c = \operatorname{ctan} A \cdot \operatorname{ctan} B = \frac{1}{\tan A \cdot \tan B} \qquad\qquad\text{(A2.6)}$$

A2.2 Theorems concerning arbitrary spherical triangles

All theorems concerning arbitrary spherical triangles can be derived from those concerning right-angled ones. An arbitrary spherical triangle is completely defined when three quantities are known. Usually, no angle is a right angle, and the sum of the angles is not 180°. In order to determine a quantity in a spherical triangle, the equation must contain four quantities, three of them known, e.g. all the angles.

The following formulas are considered most useful for spherical triangle calculations:

Sines theorem

$$\frac{\sin a}{\sin A} = \frac{\sin b}{\sin B} = \frac{\sin c}{\sin C} \qquad\qquad\text{(A2.7)}$$

Cosines theorem

$$\cos a = \cos b \cdot \cos c + \sin b \cdot \sin c \cdot \cos A \qquad (A2.8)$$

Second Cosine theorem

$$\cos A = -\cos B \cdot \cos C + \sin B \cdot \sin C \cdot \cos a \qquad (A2.9)$$

(notice the minus sign before the first member).

The cotangent theorem uses four sequentially numbered quantities, starting with a *side*:

$$\operatorname{ctan} 1 \cdot \sin 3 - \operatorname{ctan} 4 \cdot \sin 2 = \cos 2 \cdot \cos 3 \qquad (A2.10)$$

(The four quantities may be $1 = a$, $2 = C$, $3 = b$, $4 = A$.)

The sine–cosine theorem (numbering as above):

$$\sin 1 \cdot \cos 2 = -\cos 3 \cdot \cos 4 \cdot \sin 5 + \cos 5 \cdot \sin 3 \qquad (A2.11)$$

If the numbering starts with an angle, Equation (A2.11) changes into the second sine–cosine theorem:

$$\sin 1 \cdot \cos 2 = \cos 3 \cdot \cos 4 \cdot \sin 5 + \cos 5 \cdot \sin 3 \qquad (A2.12)$$

(Notice the change of the sign from Equation (A2.10) to (A2.12). The theorem contains in total five of the six quantities of the triangle. Only three of these are independent. Thus the theorem cannot be used for determining a fourth quantity from three given quantities.)

Appendix 3
The radii of curvature of an ellipsoidal surface

It may sometimes be desirable to compute the radius of curvature of the earth, e.g. to determine corrections for the movement of a gyrostabilized platform. Because of the ellipsoidal shape, such a radius depends on the direction of the curve on the earth's surface. The two main directions are along a meridian and perpendicular to it. In the first case, Figure A1.1 and Equation (A1.1) and Equation (1.1) can be utilized.

The general expression for the radius of curvature of a plane curve $y = f(x)$ is

$$R = \frac{\left[1 + \left(\dfrac{dy}{dx}\right)^2\right]^{3/2}}{\left|\dfrac{d^2 y}{dx^2}\right|} \tag{A3.1}$$

Along a meridian,

$$M = \frac{\left[1 + \left(\dfrac{dz}{dw}\right)^2\right]^{3/2}}{\left|\dfrac{d^2 z}{dw^2}\right|} \tag{A3.2}$$

$$\frac{dz}{dw} = -\frac{b^2}{a^2} \cdot \frac{w}{z} = -\operatorname{ctan} \Phi \tag{A3.3}$$

$$1 + \left(\frac{dz}{dw}\right)^2 = 1 + \operatorname{ctan}^2 \Phi = \sin^{-2} \Phi \tag{A3.4}$$

$$\frac{d^2 z}{dw^2} = -\frac{b^2}{a^2} \cdot \frac{z - w \cdot (dz/dw)}{z^2} = -\frac{b^4}{a^2 z^3} \tag{A3.5}$$

giving

$$M = \frac{a^2 z^3}{b^4 \sin^3 \Phi} = \frac{z^3}{a^2 (1 - e^2)^2 \sin^3 \Phi} \tag{A3.6}$$

From Equation (A1.5)

$$z^3 = \frac{a^3 (1 - e^2)^3 \sin^3 \Phi}{(1 - e^2 \sin^2 \Phi)^{3/2}} \tag{A3.7}$$

which gives

$$M = \frac{a(1 - e^2)}{(1 - e^2 \sin^2 \Phi)^{3/2}} \tag{A3.8}$$

The radius of curvature of the earth's surface perpendicularly to a meridian can be found by utilization of the rotational symmetry. This symmetry implies that the perpendicular N in Figure A1.1 intersects the z-axis at the same point for all points, P' having the same latitude. By definition, the length of N is then the radius of curvature. From Figure A1.1 and Equation (A1.4)

$$N = \frac{a}{\sqrt{1 - e^2 \sin^2 \Phi}} \tag{A3.9}$$

In order to illustrate the magnitude of these quantities we look at the ratio

$$\frac{M}{N} = \frac{1 - e^2}{1 - e^2 \sin^2 \Phi} < 1 \tag{A3.10}$$

Thus, the perpendicular radius of curvature always exceeds the meridional radius except at the poles where equality occurs. The ratio is at its minimum at the equator where it is 0.9933.

For obvious reasons, the radius of curvature, R, for an arbitrary azimuth must lie between M and N. It can be shown, as in Holsen (1986), that the result is

$$\frac{1}{R} = \frac{\cos^2 A}{M} + \frac{\sin^2 A}{N} \tag{A3.11}$$

where A is the azimuth of the direction for which the radius of curvature is to be calculated. Using Equations (A3.8)–(A3.9), this expression can be written

$$R = \frac{a}{\left(1 + \dfrac{e^2}{1 - e^2} \cos^2 A \cos^2 \Phi\right) \sqrt{1 - e^2 \sin^2 \Phi}} \qquad \text{(A3.12)}$$

Appendix 4
The standard deviation of the error in a two-dimensional position

For two density functions in the x- and y-directions, respectively, where $\sigma_x = \sigma_y = \sigma$ and $m_x = m_y = 0$, the joint density function for $\rho_{xy} = 0$ (i.e. perpendicular lines of position) is (Equation (2.21))

$$p(x, y)\,dx\,dy = \frac{1}{2\pi\sigma^2}\exp\left(-\frac{x^2 + y^2}{2\sigma^2}\right)dx\,dy \qquad (A4.1)$$

Transformation into polar coordinates gives

$$p(r, \varphi)\,dr\,d\varphi = \frac{r}{2\pi\sigma^2}\exp\left(-\frac{r^2}{2\sigma^2}\right)dr\,d\varphi \qquad (A4.2)$$

and by integration in φ

$$p(r)dr = \frac{r}{\sigma^2}\exp\left(-\frac{r^2}{2\sigma^2}\right)dr \qquad (A4.3)$$

The standard deviation is given by the integral

$$\sigma_r^2 = \int_0^\infty \frac{r^3}{\sigma^2}\exp\left(-\frac{r^2}{2\sigma^2}\right)dr$$

$$= \int_0^\infty -r^2\exp\left(-\frac{r^2}{2\sigma^2}\right) + \int_0^\infty 2r\exp\left(-\frac{r^2}{2\sigma^2}\right)dr = 2\sigma^2$$

$$\qquad (A4.4)$$

Thus,

$$\sigma_r = \sqrt{2}\sigma \qquad (A4.5)$$

The error circle is then, according to Equation (2.23),

$$\frac{1}{2}\left(\frac{r^2}{\sigma^2}\right) = c^2 \tag{A4.6}$$

where $r^2 = x^2 + y^2$, and σ is the standard deviation of the single line of position. Equation (2.24) gives the probability of an error circle with a given radius ($r = \sigma\sqrt{2}$ gives $p = 39.3$ per cent, $r = 2\sigma\sqrt{2}$ gives $p = 86.4$ per cent, etc.).

Appendix 5
Position determination by the method of least squares

A description is given below of the method of least squares used for position determination in a plane by means of distance measurements, but the method is general and can be applied to more than one dimension without changes. Hyperbolic systems can also be treated in the same way.

In Figure A5.1, P' with coordinates E' and N' is the assumed position and R_1 is a known position to which distance is measured and which has the coordinates E_1 and N_1. The calculated distance between P' and R_1 is then

$$C_1 = \sqrt{(E_1 - E')^2 + (N_1 - N')^2} \qquad (A5.1)$$

Differentiation gives

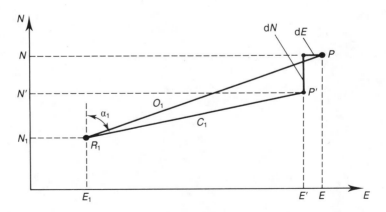

Figure A5.1 Distance measurement from an assumed position to a known reference

$$dC_1 = \frac{-1}{C_1} [(E_1 - E') dE + (N_1 - N') dN] \tag{A5.2}$$

Figure A5.1 shows that Equation (A5.2) can also be written as

$$dC_1 = \sin \alpha_1 \, dE + \cos \alpha_1 \, dN \tag{A5.3}$$

where α_1 is the directional angle from R_1 to P'. As an abbreviation we use $K_1 = \sin \alpha_1$ and $L_1 = \cos \alpha_1$. The difference between the calculated distance C_1 and the measured one O_1 should be utilized to change the assumed position P' to the (according to the measurements), correct one, P. The task is then to find the corrections dE and dN which reduce $O_1 - C_1$ to 0, i.e.

$$dC_1 + C_1 = O_1 \tag{A5.4}$$

However, there are two variables, dE and dN, to be solved in Equation (A5.4), so at least one more equation is needed, i.e. one additional measurement more to another reference point. If more reference stations are available, an equation system of the type

$$\left. \begin{array}{l} K_1 dE + L_1 dN = O_1 - C_1 \\ \vdots \\ K_n dE + L_n dN = O_n - C_n \end{array} \right\} \tag{A5.5}$$

is obtained, or, written in matrix form,

$$\mathbf{Ax} = \mathbf{b} + \mathbf{v} \tag{A5.6}$$

where

$$\mathbf{A} = \begin{bmatrix} K_1 & L_1 \\ K_2 & L_2 \\ \vdots & \\ K_n & L_n \end{bmatrix}, \; \mathbf{x} = \begin{bmatrix} dE \\ dN \end{bmatrix}, \; \mathbf{b} = \begin{bmatrix} O_1 - C_1 \\ O_2 - C_2 \\ \vdots \\ O_n - C_n \end{bmatrix}, \; \mathbf{v} = \begin{bmatrix} V_1 \\ V_2 \\ \vdots \\ V_n \end{bmatrix} \tag{A5.7}$$

$\mathbf{v}$ is the measurement random error vector. When $n > 2$, Equation (A5.6) cannot be solved exactly, but the so-called method of least squares can be utilized. In this case this method implies that the solution is those values of dE and dN which have the property of minimizing

$$\sum_{i=1}^{n} (K_i\,dE + L_i\,dN - O_i + C_i)^2$$

Thus, the method of least squares implies that the sum of squares of the deviations is minimized. If certain measurements are assumed to be less erroneous than others, more weight can be put on them, giving the function

$$\sum_{i=1}^{n} w_i(K_i\,dE + L_i\,dN - O_i + C_i)^2$$

where $0 < w_i < 1$, to be minimized with regard to dE and dN. Written in matrix form a minimum is sought of

$$f(\mathbf{x}) = (\mathbf{Ax} - \mathbf{b})^{\mathsf{T}}\mathbf{W}(\mathbf{Ax} - \mathbf{b}) \tag{A5.8}$$

where T means transposed.

Finding the minimum of $f(\mathbf{x})$ means solving the equation

$$\frac{df(\mathbf{x})}{d\mathbf{x}} = \mathbf{0} \tag{A5.9}$$

i.e.,

$$\frac{df(\mathbf{x})}{d\mathbf{x}} = \frac{d}{d\mathbf{x}}\,[(\mathbf{x}^{\mathsf{T}}\mathbf{A}^{\mathsf{T}} - \mathbf{b}^{\mathsf{T}})(\mathbf{WAx} - \mathbf{Wb})]$$

$$= \frac{d}{d\mathbf{x}}\,(\mathbf{x}^{\mathsf{T}}\mathbf{A}^{\mathsf{T}}\mathbf{WAx} - \mathbf{x}^{\mathsf{T}}\mathbf{A}^{\mathsf{T}}\mathbf{Wb} - \mathbf{b}^{\mathsf{T}}\mathbf{WAx} + \mathbf{b}^{\mathsf{T}}\mathbf{Wb})$$

$$= 2\mathbf{A}^{\mathsf{T}}\mathbf{WAx} - 2\mathbf{A}^{\mathsf{T}}\mathbf{Wb} = \mathbf{0} \tag{A5.10}$$

and consequently,

$$\mathbf{x}(f_{\min}) = (\mathbf{A}^{\mathsf{T}}\mathbf{WA})^{-1}\mathbf{A}^{\mathsf{T}}\mathbf{Wb} \tag{A5.11}$$

The solution (A5.11) normally implies that none of the equations

$$K_i\,dE + L_i\,dN = O_i - C_i$$

is satisfied exactly. This can be checked by computing the residuals

$$\mathbf{v} = \mathbf{Ax} - \mathbf{b} \tag{A5.12}$$

which also have the property that (Equations (A5.12) and (A5.11))

$$\mathbf{A}^{\mathrm{T}}\mathbf{W}\mathbf{v} = \mathbf{O} \qquad\qquad\qquad (A5.13)$$

A check of the residuals $\mathbf{v}$ shows they are all about the same size. If any one of them is considerably larger than the others, there is most probably a large error in the corresponding measurement.

The above method is not very suitable for manual calculation, but it is used in many microprocessor-controlled navigational receivers. Usually, only a few repetitions are required until the increments dE and dN become acceptably small. If the convergence is slow, this most often indicates a poor choice of starting coordinates or poorly conditioned measurements.

The weight matrix $\mathbf{W}$ can be chosen by the user, but statistically there is an optimum value. It can be shown (Gelb, 1974) that this optimum is the inverse covariance matrix (see below). In Equation (A5.12) all the N components of the vector $\mathbf{v}$ according to the assumptions have a Gaussian distribution function with mean 0. The joint density function of $\mathbf{v}$ is (Cramér, 1946)

$$P(\mathbf{v}) = (2\pi)^{-n/2}|\mathbf{R}|^{1/2} \exp\left[-\tfrac{1}{2}\boldsymbol{v}^{\mathrm{T}}\mathbf{R}^{-1}\boldsymbol{v}\right] \qquad (A5.14)$$

where

$$\mathbf{R} = E(\mathbf{v}\mathbf{v}^{\mathrm{T}}) = \begin{bmatrix} \overline{v_1^2} & \overline{v_1 v_2}\cdots & & \overline{v_1 v_n} \\ \overline{v_2 v_1} & & \overline{v_2^2} & \overline{v_2 v_n} \\ \vdots & & & \\ \overline{v_n v_1} & & & \overline{v_n^2} \end{bmatrix} \qquad (A5.15)$$

$\mathbf{R}$ is the covariance matrix of the error vector $\mathbf{v}$. It is always symmetrical. (Equation (2.21) is the two-dimensional version of Equation (A5.14).)

The expression in the exponent of Equation (A5.14) is analogous to Equation (A5.8), and obviously the density function has its maximum when $\mathbf{v}^{\mathrm{T}}\mathbf{R}^{-1}\mathbf{v} = (\mathbf{A}\mathbf{x} - \mathbf{b})^{\mathrm{T}}\mathbf{R}^{-1}(\mathbf{A}\mathbf{x} - \mathbf{b})$ has its minimum. This is the well-known 'maximum likelihood' principle from estimation theory. Obviously, this gives the same result as the method of least squares when the weighting function $\mathbf{W}$ is chosen equal to $\mathbf{R}^{-1}$. This may be regarded as an indication of the optimum choice of $\mathbf{W}$. Thus, if the measurements are uncorrelated,

$$\mathbf{W}_{\text{opt}} = \begin{bmatrix} \dfrac{1}{\sigma_1^2} & 0 & 0\ldots & & 0 \\ 0 & \dfrac{1}{\sigma_2^2} & & & 0 \\ \vdots & & & & \\ 0 & & & & \dfrac{1}{\sigma_n^2} \end{bmatrix} \qquad (A5.16)$$

This shows more clearly that for uncorrelated measurements the optimum weighting function is such that each measurement is weighted with a factor inversely proportional to the measurement error variance.

In many practical problems, quantities other than the coordinates can be inaccurate, i.e. the deviation of the receiver frequency from its nominal value. Such deviations can simply be included in the optimization routines, given that the number of measurements is greater than (or possibly equal to) the number of unknowns.

Error computation by the method of least squares can be performed in a relatively simple way. As $\mathbf{x}$ in Equation (A5.11) is a vector, the error is directionally dependent. This is expressed by the covariance matrix of the error in $\mathbf{x}$, and this matrix reflects the influence in different coordinate directions by the errors in the measurement vector $\mathbf{b}$:

$$\delta\mathbf{x} = (\mathbf{A}^T\mathbf{W}\mathbf{A})^{-1}\mathbf{A}^T\mathbf{W}\delta\mathbf{b} \qquad (A5.17)$$

$$C_{\text{ov}}(\delta\mathbf{x}) = E\{(\mathbf{A}^T\mathbf{W}\mathbf{A})^{-1}\mathbf{A}^T\mathbf{W}\delta\mathbf{b}[(\mathbf{A}^T\mathbf{W}\mathbf{A})^{-1}\mathbf{A}^T\mathbf{W}\delta\mathbf{b}]^T\}$$

$$= (\mathbf{A}^T\mathbf{W}\mathbf{A})^{-1}\mathbf{A}^T\mathbf{W}\mathbf{R}\mathbf{W}^T\mathbf{A}[(\mathbf{A}^T\mathbf{W}\mathbf{A})^{-1}]^T \qquad (A5.18)$$

where

$$\mathbf{R} = C_{\text{ov}}(\mathbf{v}) = E(\delta\mathbf{b}\,\delta\mathbf{b}^T) \qquad (A5.19)$$

As $\mathbf{W} = \mathbf{R}^{-1}$ (whose symmetry also makes $(\mathbf{A}^T\mathbf{W}\mathbf{A})^{-1}$ symmetric), Equation (A5.18) can be simplified to

$$\text{Cov}(\delta\mathbf{x}) = (\mathbf{A}^T\mathbf{W}\mathbf{A})^{-1} = \begin{bmatrix} \sigma_{11}^2 & \sigma_{21}^2\ldots & & \sigma_{n1}^2 \\ \sigma_{21}^2 & \sigma_{22}^2 & & \sigma_{n2} \\ \vdots & & & \\ \sigma_{n1}^2 & & & \sigma_{nn}^2 \end{bmatrix} \qquad (A5.20)$$

where σ_{ii} is the standard deviation in the ith coordinate direction, and σ_{ij} is the correlation between the errors in the ith and jth directions. The rms non-directional error becomes

$$\sigma^2 = \sum_{i=1}^{n} \sigma_{ii}^2 \qquad\qquad \text{(A5.21)}$$

Equation (A5.20) can also be utilized for computations of error ellipses when $n = 2$, using Equations (2.25), (2.32)–(2.35) and (2.39)–(2.40) with

$$x_{11} = \sigma_{11}^{-2} \qquad\qquad \text{(A5.22)}$$

$$x_{12} = -2\sigma_{12}^2 \sigma_{11}^{-2} \sigma_{22}^{-2} \qquad\qquad \text{(A5.23)}$$

$$x_{22} = \sigma_{22}^{-2} \qquad\qquad \text{(A5.24)}$$

$$k = -2(1 - \sigma_{12}^4 \sigma_{11}^{-2} \sigma_{22}^{-2}) \ln(1 - p) \qquad\qquad \text{(A5.25)}$$

Appendix 6
Errors in DECCA user lines of position due to phase errors caused by the master-to-slave, master-to-user and slave-to-user signal paths

A vector diagram showing how phase errors arise is given in Figure A6.1. At a complete position measurement there are phase errors in both the synchronization between the master and the slave and in the receiver, so that the total interference picture is as shown in Figure A6.2. Here the receiver has been marked with X, the master with M, and the red, green, and purple slaves with R, G and P, respectively,

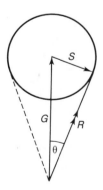

Figure A6.1 Vector diagram showing the resulting phase error θ when the sky wave S with arbitrary phase interferes with the ground wave G so that the received signal is R

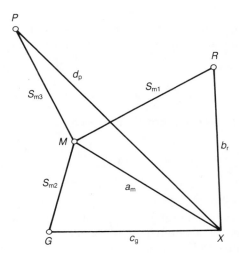

Figure A6.2 Phase deviation along the paths of transmission of the red, green and purple signals

S_{m_1} is the phase error, expressed in periods at the master frequency, of the synchronizing signal from the red slave, S_{m_2} is the corresponding error of the synchronization of the green slave, etc., a_m is the phase error in periods at the master frequency of the signal received from the master by the receiver, b_r is the phase error in periods of the red frequency of the received signal from the red slave, etc.

The red slave transmits a frequency which is $\frac{4}{3}$ of the master frequency. The resulting phase error in the red signal (in periods of the red frequency) at the receiver is, consequently, $b_r + 4S_{m_1}/3$. In the navigation receiver the red signal is multiplied by 3 and the master signal by 4 before the phase detection so that the resulting error of the phase difference at the measurement frequency is

$$\theta_R = 4a_m - 3(b_r + 4S_{m_1}/3) \tag{A6.1}$$

Here the error is expressed in periods at the measurement frequency $24f$.

The corresponding equations of the green and purple signals are (Figure A6.2).

$$\theta_G = 3a_m - 2(c_g + S_{m_2}/2) \tag{A6.2}$$

$$\theta_P = 4a_m - 6(d_P + 5S_{m_3}/6) \tag{A6.3}$$

In order to find the standard deviation of the error of the red phase measurement the squared result of a large number of measurements is added and the sum is divided by the number of measurements.

$$\sigma_{RN}^2 = \frac{1}{N} \sum_{n=1}^{N} \theta_{Rn}^2 \tag{A6.4}$$

Equation (A6.1) inserted into Equation (A6.4) gives

$$\sigma_{RN}^2 = \frac{1}{N} \sum_{n=1}^{N} (4a_{mn} - 3b_{rn} - 4S_{m_1n})^2 \tag{A6.5}$$

The phase errors of the different transmission paths are independent random variables having the same probability of positive and negative values. The sum of a number of products of errors of two different signals thus approaches zero when the number is very large. For large values of N, Equation (A6.5) can thus be written

$$\sigma_R^2 = \lim_{N \to \infty} \sigma_{RN}^2 \tag{A6.6}$$

$$\sigma_R^2 = 16\sigma_{mn}^2 + 9\sigma_{br}^2 + 16\sigma_{sm_1}^2 \tag{A6.7}$$

The meaning of Equation (A6.5) is that if a large number of red phase measurements are carried out at a fixed site in the presence of a skywave, an error occurs, the average of which is zero and the variance of which can be written as a weighted sum of the variances of the phase errors of each transmission path.

The sum in Equation (A6.7) should be added to the variance of the possible phase error caused by the phase-locking of the slave signal to the master signal (in the slave transmitter). The variance is 0.012^2 periods of the transmitted frequency but independent of the sky wave (DECCA, 1979). Some slave transmitters which suffer particularly from sky wave interferene in the master signal use a rubidium clock as a reference at night, replacing the third term of Equation (4.17) by the variance of the phase noise of the clock signal.

(The resulting phase error also contains errors of the phase measurement itself because of receiver noise, but this has not been considered here, as only errors caused by the transmission path have been taken into consideration.)

Using the same procedure as above the variances of the errors of the green and purple phase measurements can be derived. From Equations (4.12) and (4.13) we obtain

$$\sigma_G^2 = 9\sigma_{am}^2 + 4\sigma_{cg}^2 + 9\sigma_{sm_2}^2 \tag{A6.8}$$

$$\sigma_P^2 = 25\sigma_{am}^2 + 36\sigma_{dP}^2 + 25\sigma_{sm_3}^2 \tag{A6.9}$$

If the variance of the phase-lock errors of the slaves is also con-sidered, σ_G^2 is increased by 0.009^2 and σ_p^2 by 0.015^2 (DECCA, 1979).

Table A6.1 gives a few measured valid values for summer nights at temperate latitudes as an example of the magnitude of the parameters of Equations (A6.7)–(A6.9).

Table A6.2 gives typical red lines-of-position accuracies (1σ) on the front of a chain with a baseline of 160 km.

Table A6.1 Standard deviation σ_τ in periods of transmitted frequency for insertion into Equations (A6.7)–(A6.9)

| Range (km) | σ_T in transmitted cycles for good soil and sea water | | | |
	a_m S_{m_1} S_{m_2} S_{m_3}	b_r	c_g	d_p
100	0.0041	0.0041	0.0041	0.0050
200	0.0100	0.0100	0.0100	0.0118
300	0.0187	0.0187	0.0187	0.0224
400	0.0335	0.0335	0.0335	0.0400
500	0.0532	0.0532	0.0532	0.0632
Range (km)	σ_T in transmitted cycles for poor soil			
100	0.0044	0.0053	0.0053	0.0053
200	0.0106	0.0053	0.0053	0.0125
300	0.0224	0.0266	0.0266	0.0266
400	0.0400	0.0472	0.0472	0.0472
500	0.0707	0.0845	0.0845	0.0845

Source: Decca, 1979. Courtesy Racal–Decca Marine Navigation Ltd.

Table A6.2 Typical accuracies of the red line of position in lane widths and meters 68% probability, 160 km baseline

| Range (km) | Summer day | | Winter day | | Summer night | | Winter night | |
	lane	m	lane	m	lane	m	lane	m
100	0.010	7	0.014	12	0.028	22	0.040	30
200	0.013	18	0.028	40	0.056	76	0.080	100
300	0.015	30	0.047	90	0.095	170	0.140	240
400	0.020	55	0.080	200	0.160	400	0.240	550
500	0.030	100	0.120	380	0.250	750	0.350	1050

Source: Decca, 1979. Courtesy Racal–Decca Marine Navigation Ltd.

Appendix 7
Satellites in orbit

A7.1 Derivation of Kepler's laws

(These laws were formulated informally in Section 7.1.)

Starting with Equation (7.1) and Newton's second law, which states that the force is equal to the mass times the acceleration, we have

$$ma = -m\mu/r^2 \tag{A7.1}$$

(the minus sign indicates that the power is in the negative r-direction). Because of the radial direction of the force, polar coordinates are suitable. The common relation between Cartesian and polar coordinates is (Figure A1.1)

$$x = r\cos\Phi \tag{A7.2}$$
$$y = r\sin\Phi \tag{A7.3}$$

After differentiation with regard to time, the corresponding velocities are

$$\dot{x} = \dot{r}\cos\Phi - r\dot{\Phi}\sin\Phi \tag{A7.4}$$
$$\dot{y} = \dot{r}\sin\Phi + r\dot{\Phi}\cos\Phi \tag{A7.5}$$

and after another time differentiation the accelerations are

$$\ddot{x} = \ddot{r}\cos\Phi - 2\dot{r}\dot{\Phi}\sin\Phi - r\ddot{\Phi}\sin\Phi - r\dot{\Phi}^2\cos\Phi \tag{A7.6}$$
$$\ddot{y} = \ddot{r}\sin\Phi + 2\dot{r}\dot{\Phi}\cos\Phi + r\ddot{\Phi}\cos\Phi - r\dot{\Phi}^2\sin\Phi \tag{A7.7}$$

The acceleration in polar coordinates has two orthogonal components a_r and a_Φ. For these, Equations (A7.2) and (A7.3) must apply, i.e.

$$\ddot{x} = a_r \cos\Phi - a_\Phi \sin\Phi \qquad (A7.8)$$

$$\ddot{y} = a_r \sin\Phi + a_\Phi \cos\Phi \qquad (A7.9)$$

This gives

$$a_r = \ddot{r} - r\dot{\Phi}^2 \qquad (A7.10)$$

$$a_\Phi = 2\dot{r}\dot{\Phi} + r\ddot{\Phi} \qquad (A7.11)$$

According to Equation (A7.1), $a_r = -\mu/r^2$. There is no force orthogonal to this, so

$$\ddot{r} - r\dot{\Phi}^2 = -\mu/r^2 \qquad (A7.12)$$

$$2\dot{r}\dot{\Phi} + r\ddot{\Phi} = 0 \qquad (A7.13)$$

Equation (A7.13) can be written

$$\frac{1}{r}\frac{d}{dt}(r^2\dot{\Phi}) = 0 \qquad (A7.14)$$

and, consequently,

$$r^2\dot{\Phi} = \text{const.} = c_1 \qquad (A7.15)$$

Kepler's second law is illustrated in Figure A7.1. (Notice that the only assumption behind this result is Equation (A7.11), i.e. that earth's centre of gravitation is the only external force.)

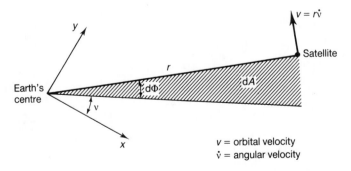

Figure A7.1 The radius vector of the satellite sweeps over equal areas per unit of time

From Equation (A7.15) and Figure A7.1, we have

$$dA = \tfrac{1}{2}r(r\dot{\Phi})\,dt = \tfrac{1}{2}c_1\,dt \qquad\qquad (A7.16)$$

and, after integration,

$$A = \tfrac{1}{2}c_1(t_2 - t_1) \qquad\qquad (A7.17)$$

In order to derive Kepler's first law we rewrite Equation (A7.15) as

$$r\dot{\Phi} = c_1/r \qquad\qquad (A7.15a)$$

After differentation with regard to Φ:

$$\dot{\Phi}\frac{dr}{d\Phi} + \frac{d\dot{\Phi}}{d\Phi}\,r = -\frac{c_1}{r^2}\cdot\frac{dr}{d\Phi} = -\frac{c_1}{r^2}\cdot\frac{\dot{r}}{\dot{\Phi}} \qquad\qquad (A7.18)$$

But

$$\frac{d\dot{\Phi}}{d\Phi}\,r = \frac{\ddot{\Phi}}{\dot{\Phi}}\,r = -2\dot{r} \qquad\qquad (A7.19)$$

(see Equation (A7.13). Thus Equation (7.18) can be written

$$\dot{\Phi}\frac{dr}{d\Phi} = 2\dot{r} - \frac{c_1\dot{r}}{r^2\dot{\Phi}} = \dot{r} \qquad\qquad (A7.20)$$

(see Equation (A7.15a)), i.e.

$$\dot{r} = \dot{\Phi}\frac{dr}{d\Phi} = \frac{c_1\dot{r}}{r^2\dot{\Phi}} = \frac{c_1}{r^2}\cdot\frac{dr}{d\Phi} = -\frac{d}{d\Phi}\left(\frac{c_1}{r}\right) \qquad\qquad (A7.21)$$

Time differentiation of Equation (A7.21) and insertion of Equation (A7.15) give

$$\ddot{r} = -\frac{d^2}{d\Phi^2}\left(\frac{c_1}{r}\right)\dot{\Phi} = -\frac{c_1}{r^2}\frac{d^2}{d\Phi^2}\left(\frac{c_1}{r}\right) \qquad\qquad (A7.22)$$

Inserting Equations (A7.12) and (A7.15) into Equation (A7.22):

$$-\frac{c_1}{r^2}\cdot\frac{d^2}{d\Phi^2}\left(\frac{c_1}{r}\right) - \left(\frac{c_1}{r}\right)^2\!/r = -\mu/r^2 \qquad\qquad (A7.23)$$

which can be rewritten

$$\frac{d^2}{d\Phi^2}\left(\frac{c_1}{r}\right) + \frac{c_1}{r} = \frac{\mu}{c_1}$$

(A7.24)

By regarding c_1/r as the dependent variable, the solution of the differential equation is

$$\frac{c_1}{r} = A \sin \Phi + B \cos \Phi + \frac{\mu}{c_1}$$

(A7.24a)

where A and B are constants. Thus

$$r = \frac{c_1}{\dfrac{\mu}{c_1} + A \sin \Phi + B \cos \Phi}$$

(7.25)

Any sum of sine and cosine functions can be written as another cosine function with different phase and amplitude, such that

$$A \sin \Phi + B \cos \Phi = \sqrt{A^2 + B^2} \cos\left(\arctan \frac{A \cos \Phi - B \sin \Phi}{A \sin \Phi + B \cos \Phi}\right)$$

(A7.26)

Consequently, Equation (A7.25) can be rewritten as

$$r = \frac{c_1^2/\mu}{1 + \dfrac{c_1}{\mu}\sqrt{A^2 + B^2} \cos\left(\arctan\dfrac{A \cos \Phi - B \sin \Phi}{A \sin \Phi + B \cos \Phi}\right)}$$

$$= \frac{p}{1 + e \cos v}$$

(A7.27)

Thus Kepler's first law has been proved as Equation (A7.27) is the equation of a conical section in polar coordinates where the angle v is zero at perigee (Figure A7.2).

The eccentricities of the ellipse, the parabola and the hyperbola are <1, 1 and >1 respectively. The parameter p is the positive y-coordinate of the curve when the x-coordinates is that of the focus (Figure A7.2). Equation (A7.27) can be derived for the three types of intersection by transformation of the equations of the curves from Cartesian to polar coordinates, using the relations $r \cos v = x - ae$ and $r \sin v = y$ (Figure A7.2).

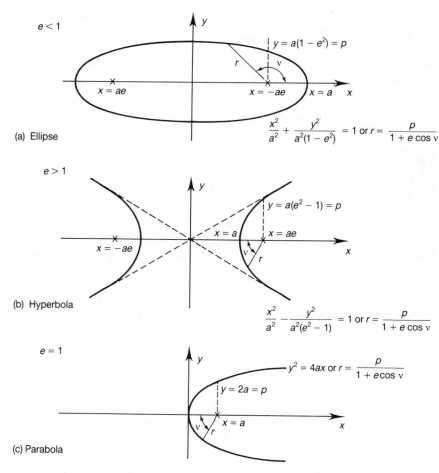

Figure A7.2 The conical intersections

Which of the three possibilities Equation (A7.27) describes is determined by the parameters p and e (i.e. by the integration constants A and B). If the velocity at perigee, v_p, is known,

$$r_p \dot{v}_p = v_p \tag{A7.28}$$

and, according to Equations (A7.15) and (A7.27),

$$r_p v_p = r_p^2 \dot{v}_p = c_1 = \sqrt{\mu p} \tag{A7.29}$$

i.e.

$$p = r_p^2 v_p^2/\mu \tag{A7.30}$$

When $v = 0$, $r = r_p$, so that

$$r_p = \frac{p}{1 + e} = \frac{r_p^2 v_p^2/\mu}{1 + e} \tag{A7.31}$$

i.e.

$$e = \frac{r_p v_p^2}{\mu} - 1 \tag{A7.32}$$

Thus, the orbit is completely defined if the velocity and the distance at perigee are given.

Now, Kepler's third law can be derived from Equation (A7.17). A satellite in elliptical orbit around the earth has the orbital time T, i.e.

$$\pi a^2 \sqrt{1 - e^2} = A_e = T c_1/2 \tag{A7.33}$$

According to Equations (A7.29) and Figure A7.2(a),

$$c_1 = \sqrt{\mu p} = \sqrt{\mu a(1 - e^2)} \tag{A7.34}$$

so that

$$T^2 = \frac{4\pi^2}{\mu} a^3 \tag{A7.35}$$

which is Kepler's third law.

A7.2 The orbital velocity

The orbital velocity (Equations (A7.4), (A7.5)) is given as

$$v = \sqrt{\dot{x}^2 + \dot{y}^2} = \sqrt{\dot{r}^2 + (r\dot{v})^2} \tag{A7.36}$$

and the radial velocity is (Equations (A7.25) and (A7.27))

$$\dot{r} = \frac{dr}{dv}\dot{v} = \frac{pe \sin v}{(1 + e \cos v)^2}\dot{v} = \frac{c_1}{r^2}\frac{pe \sin v}{(1 + e \cos v)^2}$$

$$= \frac{\sqrt{\mu p}\, pe \sin v}{r^2(1 + e \cos v)^2} = \sqrt{\frac{\mu}{p}}\, e \sin v \tag{A7.37}$$

Further, we have

$$(r\dot v)^2 = (r^2\dot v)^2/r^2 = \frac{c_1^2}{r^2} = \frac{\mu p}{p^2/(1 + e\cos v)^2}$$

$$= \frac{\mu}{p}(1 + e\cos v)^2 \tag{A7.38}$$

Equation (A7.36) gives

$$v^2 = \frac{\mu}{p}[e^2\sin^2 v + (1 + e\cos v)^2] = \frac{\mu}{p}\left[e^2(1 - \cos^2 v) + \left(\frac{p}{r}\right)^2\right]$$

$$= \frac{\mu}{p}\left[e^2\left(1 - \left(\frac{\frac{p}{r} - 1}{e}\right)^2\right) + \left(\frac{p}{r}\right)^2\right]$$

$$= \frac{\mu}{p}\left(e^2 - 1 + \frac{2p}{r}\right) = \mu\left(\frac{2}{r} - \frac{1 - e^2}{p}\right) \tag{A7.39}$$

For an elliptic orbit $p = a(1 - e^2)$ and, consequently,

$$v = \sqrt{\mu\left(\frac{2}{r} - \frac{1}{a}\right)} \tag{A7.40}$$

At the perigee the velocity is v_p and the focal distance $r_p = a(1 - e)$ so that

$$v_p = \sqrt{\frac{\mu}{a} \cdot \frac{1 + e}{1 - e}} \tag{A7.41}$$

In the limit when a satellite leaves an earth orbit, the orbit is a parabola, i.e. $a = \infty$, and the velocity, according to Equation (A7.40), is then

$$v_{pa} = \sqrt{\frac{2\mu}{r}} \tag{A7.42}$$

At the other limit the orbit is a circle, i.e. $a = r$, and

$$v_{pc} = \sqrt{\mu/r} \tag{A7.43}$$

The latter velocity can also be achieved by setting $e = 0$. The velocity needed in a launch parallel to the earth's surface to put a satellite into a circular orbit is, with known values of μ and of the earth radius, $\sqrt{398{,}601/6378} = 7.9$ km/s. This is called the first cosmic velocity.

The necessary velocity in a parabolic orbit can also be calculated by setting $e = 1$, which gives the same result as Equation (A7.42), i.e. $v_{pa} = \sqrt{2\mu/r_p}$. With the same values as above, this velocity is 11.2 km/s which is the second cosmic velocity.

Figure A7.3 shows an imagined situation where a mass particle is launched in parallel with the earth's surface at different velocities, v. The earth is assumed to be spherical with an homogeneous mass distribution and with no atmosphere. In the limit $v = 0$ the particle falls directly towards the earth's centre. By increasing v first, we obtain extruded elliptic orbits which, when v approaches the first cosmic velocity, obtain decreasing deviation from circular shape (decreasing e, increasing a). When v is increased further, we obtain ellipses with increasing eccentricity (increasing e, increasing a) until the eccentricity reaches unity and the particle leaves the earth's orbit.

A7.3 Orbital coordinates as time functions

The position of the satellite in polar coordinates is (Equation (A7.27)

$$r = \frac{p}{1 + e \cos v} = \frac{a(1 - e^2)}{1 + e \cos v} \qquad (A7.44)$$

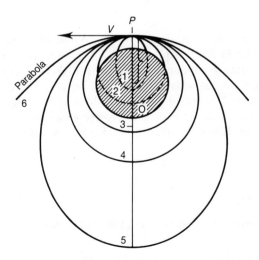

Figure A7.3 The influence of different launch velocities, (the numbers 1–6 denote increasing v)

With knowledge of the orbit, the position of the satellite as a function of time can also be calculated. From Kepler's second law (Equation (A7.15)) we obtain

$$\dot{v} = \sqrt{\mu p}/r^2 = \frac{\sqrt{\mu p}\,(1 + e\cos v)^2}{p^2}$$

$$= \sqrt{\frac{\mu}{a^3(1 - e^2)^3}}\,(1 + e\cos v)^2 \tag{A7.45}$$

i.e.

$$\frac{dv}{(1 + e\cos v)^2} = \sqrt{\frac{\mu}{a^3(1 - e^2)^3}}\,dt \tag{A7.46}$$

which, after direct integration, gives

$$\frac{1}{1 - e^2}\left[\frac{2}{\sqrt{1 - e^2}}\arctan\left(\sqrt{\frac{1 - e}{1 + e}}\tan\frac{v}{2}\right) - \frac{e\sin v}{1 + e\cos v}\right]$$

$$= \sqrt{\frac{\mu}{a^3(1 - e^2)^3}}\,(t - t_p) \tag{A7.47}$$

or

$$t - t_p = \sqrt{\frac{a^3}{\mu}}\left[2\arctan\left(\sqrt{\frac{1 - e}{1 + e}}\tan\frac{v}{2}\right) - \frac{e\sqrt{1 - e^2}\sin v}{1 + e\cos v}\right] \tag{A7.47a}$$

where t_p is the time of perigee ($v = 0$). When the relation between t and v is calculated from Equation (A7.47), the distance r is given as a function of time by means of Equation (A7.44). Equation (A7.47) shows that the true anomaly v can only be calculated implicitly as a function of time. For navigational satellites with orbits having $e \ll 1$, it is practical to utilize an average velocity which is then a constant.

According to Kepler's third law (Equation (A7.35)), the orbital period is

$$T = 2\pi\sqrt{\frac{\mu}{a^3}} \tag{A7.48}$$

and, consequently, the average angular velocity is

$$\eta = \frac{2\pi}{T} = \sqrt{\frac{\mu}{a^3}} \tag{A7.49}$$

The radial velocity is given by Equation (A7.37). After inserting the equation of the ellipse (Equation (A7.44)) we obtain

$$\dot{r}^2 = \frac{\mu}{p}\left[e^2 - \left(\frac{p}{r} - 1\right)^2\right] = \mu\left(\frac{2r - p}{r^2} - \frac{1 - e^2}{p}\right)$$

$$= \mu\,\frac{a^2 e^2 - (a - r)^2}{ar^2} \tag{A7.50}$$

After inserting Equation (A7.49),

$$\eta\,dt = \frac{r\,dr}{a\sqrt{a^2 e^2 - (a - r)^2}} \tag{A7.51}$$

Equation (A7.51) can be integrated directly, but the result is more illustrative if we make the substitution

$$r = a(1 - e\cos E) \tag{A7.52}$$

A geometrical interpretation of this is shown by Figure A7.4. The angle E is called the eccentric anomaly.

From the diagram,

$$\overline{OR} = r\cos v = a\cos E - ae \tag{A7.53}$$

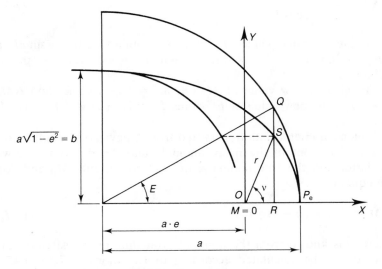

Figure A7.4 Eccentric anomaly

and from Equation (A7.44),

$$r \cos v = [a(1 - e^2) - r]/e \tag{A7.54}$$

Equations (A7.53) and (A7.54) together give Equation (A7.52). Equation (A7.51) can now be integrated with regard to E:

$$\eta \int_{t_p}^{t} dt_1 = \int_0^E \frac{(1 - e \cos E_0)(ae \sin E_0)\, dE_0}{\sqrt{a^2 e^2 - a^2 e^2 \cos^2 E_0}}$$

$$= \int_0^E (1 - e \cos E_0)\, dE_0 = E - e \sin E$$

$$= \eta(t - t_p) = M \tag{A7.55}$$

Equation (A7.55) is Kepler's equation and M is the mean anomaly.

The data signals from the navigational satellites usually contain η, t and t_p so that M can be determined. This must be done by iteration. Because of small values of e, $E = M$ is a suitable starting value. Differentiation of Kepler's equation gives

$$\Delta E(1 - e \cos E) = \Delta M \tag{A7.56}$$

or

$$\Delta E = \frac{\Delta M}{1 - e \cos E} \tag{A7.57}$$

With the starting value $E_1 = M$ we obtain $\Delta M = e \sin M$ and $\Delta E_1 = (e \sin M)/(1 - e \cos M)$, so that $E_2 = E_1 + \Delta E_1$ is the new value of E.

This gives a new value of M, i.e. $M_2 = E_2 - e \sin E_2$ and $\Delta M_2 = M - M_2$. The new value of E is thus $E_3 = E_2 + \Delta M_2/(1 - e \cos E_2)$, etc.

Kepler parameter data transmitted from a navigation satellite to the user usually contain the mean anomaly and the difference between the latter and the true anomaly, i.e. the coefficients M_0 and $\Delta \eta$ of the equation

$$M = M_0 + (\eta + \Delta \eta)t \tag{A7.58}$$

where η is known from the nominal orbit time of the satellite. From this, E can be calculated according to Equations (A7.55)–(A7.57) (the value of e is also transmitted by the satellite). Insertion of

Equation (A7.44) into Equation (A7.54) and using Equation (A7.53) then gives

$$\cos v = \frac{\cos E - e}{1 - e \cos E} \tag{A7.59}$$

After calculation of v, the receiver also calculates $r(v)$ according to Equation (A7.44). When r, v and ω (also transmitted by the satellite data signal) are known, x' and y' (Figure 7.5) can be calculated:

$$x' = r \cos (v + \omega) \tag{A7.60}$$

$$y' = r \sin (v + \omega) \tag{A7.61}$$

For position calculations, it is necessary to express the positions of the satellite and of the user in the same coordinate system, i.e. (usually) longitude, latitude and altitude (above the ellipsoid). This necessitates a transformation of the coordinates x' and y', which can be performed by first rotating the $x'-y'$-plane around the x'-axis by the inclination angle i to the equatorial plane of the earth. Then the coordinate system is rotated in this new plane around the z-axis (the rotational axis of the earth) until the x'- and x-axes coincide (through the 0 meridian) (Figure A7.5).

The relations are

$$x = x' \cos (\Omega - \Omega') - y' \cos i \sin (\Omega - \Omega') \tag{A7.62}$$

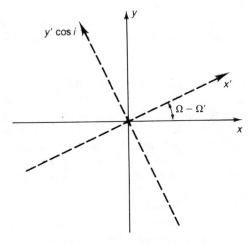

Figure A7.5 Rotation of the coordinate system

$$y = x' \sin(\Omega - \Omega') + y' \cos i \cos(\Omega - \Omega') \qquad \text{(A7.63)}$$

$$z = y' \sin i \qquad \text{(A7.64)}$$

Here, the angle of rotation, $\Omega - \Omega'$, is the difference between the rectascension, Ω, of the ascending node of the satellite orbit, and the rectascension, Ω', of the 0 meridian. As the rotational velocity of the earth is known, Ω' can be calculated as a function of time.

After calculation of x, y and z, Φ, Λ and h of the satellite can be calculated for a given ellipsoid according to Equations (A1.12)–(A1.16).

A7.4 Launching of satellites into a desired orbit

The launching of satellites into a desired orbit is difficult and very costly. The price for this today is of the order of $50,000,000. (This does not include the cost of the satellite itself.) According to previous plans, the GPS Block II production satellites were to be launched into orbit by means of the space shuttle, whereas earlier satellites (TRANSIT included) were launched by rockets. Delta rockets are now used because of the capacity problems following the shuttle disaster in January 1986. The Proton rocket is used in the Soviet Union, and NAVSAT satellites will probably be launched by the *Ariane* rocket. A common factor of all these methods is that the satellite is first brought up to a so-called transfer orbit which is strongly elliptic and where the apogee is in the designed circular orbit.

(All navigational satellites so far have used circular orbits, but in principle this is not necessary. For example, NAVSAT will also use satellites in elliptical orbits.) The satellite is in a transfer orbit for some hours, while it is tested and orientated. At the apogee it has a velocity smaller than needed to enter a circular orbit at that altitude. Usually, a so-called apogee motor is utilized to give the necessary velocity increase. This is a solid fuel motor ignited at the apogee and has to give the desired orbit inclination. There may also be several steps in the changes of the orbit from the satellite leaving its rocket (shuttle) until it reaches its final orbit.

After the satellite has entered the desired orbit, it is influenced by external forces acting to change the orbit, thus on-board fuel is necessary to counter these forces. Orbit corrections are fairly rare, but their fuel consumption often determines the length of the satellite life. High-altitude satellites, e.g. GPS, have a lifespan of 8–10 years, sometimes even more, whereas low-orbit satellites, like TRANSIT, usually have shorter lifetimes if the deviations from the desired orbit are to be kept within strict limits.

It was shown in Section A7.2 that the orbital velocity of a satellite is (Equation (A7.40))

$$v_s = \sqrt{\mu\left(\frac{2}{r} - \frac{1}{a}\right)} \tag{A7.65}$$

where r is the distance to the centre of gravity of the earth, and a is the length of the semi-major axis. As the mechanical energy of a satellite consists partly of potential energy, expressed by Mgr, and partly of kinetic energy, expressed by $0.5\,mv_s^2$, it is relatively simple to calculate the additional energy needed to change a satellite orbit. At the apogee the velocity according to Equation (A7.65) is

$$v_{sa} = \sqrt{\mu\left[\frac{2}{a(1 + e)} - \frac{1}{a}\right]} = \sqrt{\mu\frac{1 - e}{a(1 + e)}} \tag{A7.66}$$

whereas the velocity in a circular orbit at the same altitude has to be $(r = a(1 + e))$

$$v_{ss} = \sqrt{\frac{\mu}{a(1 + e)}} \tag{A7.67}$$

The additional velocity is then

$$\Delta v = v_{ss} - v_{sa} = \sqrt{\frac{\mu}{a(1 + e)}}\,(1 - \sqrt{1 - e}\,) \tag{A7.68}$$

and the necessary additional energy to achieve this is

$$\Delta E = \frac{m_s}{2}\,(v_{ss}^2 - v_{sa}^2) = \frac{m_s}{2} \cdot \frac{\mu e}{a(1 + e)} \tag{A7.69}$$

where m_s is the mass of the satellite.

The relation between Δv, the fuel velocity v_d, m_s, the mass of the empty missile m_r and the mass of the fuel m_d is (Berlin, 1988)

$$1 + \frac{m_d}{m_s + m_r} = \exp\left(\Delta v/v_d\right) \tag{A7.70}$$

and the relation between the mass of the fuel and the mass of the satellite is

$$\frac{m_d}{m_s} = \frac{\exp(\Delta v/v_d) - 1}{1 - \dfrac{K}{K + 1}\exp(\Delta v/v_d)} \tag{A7.71}$$

where $K = m_r/(m_s + m_d)$. The ratio m_d/m_s obviously approaches infinity when the second term of the nominator approaches unity.

Appendix 8
A method of satellite selection in time measurement systems

A minimum value of GDOP can be found by using Equation (9.62), together with Equations (9.61) and (9.63). The conditions in Equation (9.60) are taken into consideration by means of Lagrange multiplicators m_2, m_3 and m_4. This gives

$$P = \frac{H}{|\mathbf{G}_u|^2} + m_2(e_{2x}^2 + e_{2y}^2 - 1) + m_3(e_{3x}^2 + e_{3y}^2 + e_{3z}^2 - 1)$$
$$+ m_4(e_{4x}^2 + e_{4y}^2 + e_{4z}^2 - 1) \tag{A8.1}$$

Below, differentiation with regard to the variable is marked with the variable as subscript. The partial derivatives of P can then be written:

$$P_i = \frac{H_i|\mathbf{G}_u| - 2H|\mathbf{G}_u|_i}{|\mathbf{G}_u|^3} + 2im_j = 0 \tag{A8.2}$$

where

$i = e_{2x}, e_{3x}, e_{4x}, e_{2y}, e_{3y}, e_{4y}, e_{3z}, e_{4z}$

$j = 2$ if $i = e_{2x}$ or e_{2y}

$j = 3$ if $i = e_{3x}$ or e_{3y} or e_{3z}

$j = 4$ if $i = e_{4x}$ or e_{4y} or e_{4z}

Eliminating m_2, m_3 and m_4 in the set of Equations (A8.2) gives

$$(e_{ky}H_{e_{kx}} - e_{kx}H_{e_{ky}})|\mathbf{G}_u| - 2H[e_{ky}|\mathbf{G}_u|_{e_{kx}} - e_{kx}|\mathbf{G}_u|e_{ky}] = 0 \tag{A8.3}$$

where $k = 2, 3, 4$, and

$$(e_{nz}H_{e_{nx}} - e_{nx}H_{e_{nz}})|\mathbf{G}_u| - 2H[e_{nz}|\mathbf{G}_u|_{e_{nx}} - {}_{e_{nx}}|\mathbf{G}_u|_{e_{nz}}] = 0 \qquad \text{(A8.4)}$$

where $n = 3, 4$.

Inserting Equations (9.60) and (A8.3) into Equation (A8.4) gives

$$(\mathbf{G}_u)_{\text{opt}} = \begin{bmatrix} 1 & 0 & 0 & -1 \\ -\dfrac{1}{3} & \dfrac{2\sqrt{2}}{3} & 0 & -1 \\ -\dfrac{1}{3} & -\dfrac{\sqrt{2}}{3} & \dfrac{\sqrt{6}}{3} & -1 \\ -\dfrac{1}{3} & -\dfrac{\sqrt{2}}{3} & -\dfrac{\sqrt{6}}{3} & -1 \end{bmatrix} \qquad \text{(A8.5)}$$

As $\mathbf{G}_u$ is optimized without consideration of the elevation angles to the satellites, Equation (A8.6) gives, as expected, a symmetric satellite configuration. The four unity vectors whose components form the rows of $\mathbf{G}_u$ (except for the last column) all have the same angles, Φ, to each other, and

$$\Phi = \arccos\left(-\frac{1}{3}\right) = 109.47° \qquad \text{(A8.6)}$$

Equation (A8.5) inserted into Equation (9.62) gives

$$(GDOP)_{\min} = 1.5811 \qquad \text{(A8.7)}$$

A further examination shows that Equation (A8.5) implies that the partial derivatives of the determinant of $\mathbf{G}_u$ are 0, and this actually means that the volume of the tetrahedron is maximized.

It is unlikely that four satellites satisfying Equation (A8.5) will be found in practice. In addition, there is a requirement that the elevation angle of a satellite exceed a certain minimum. For this reason it is desirable to be able to minimize the GDOP of an available satellite constellation. As shown, GDOP is, in the first approximation, proportional to V^{-1} (Equations (9.56), (9.58) and (9.62)). It is therefore natural, and economic in regard to computer time, to select the constellation that gives the largest V. To make the solution practically usable the following assumptions are made:

- The directions to three satellites are known (in Figure 9.4 this

implies that e_{2x}, e_{2y}, e_{3x}, e_{3y} and e_{3z} are known).
• The 3rd and 4th satellites are symmetric with regard to the first one (in Figure 9.4 $e_{3x} = e_{4x}$).

Equation (9.61) can then be written

$$|G_u| = (e_{4y}e_{3z} - e_{3y}e_{4z})\,(1 - e_{2x}) + e_{2y}(1 - e_{3x})(e_{4z} - e_{3z}) = 6V \tag{A8.8}$$

The additional conditions (9.39) are still valid so that the function

$$P = (e_{4y}e_{3z} - e_{3y}e_{4z})\,(1 - e_{2x}) + e_{2y}(1 - e_{3x})\,(e_{4z} - e_{3z})$$
$$+ m(e_{4x}^2 + e_{4y}^2 + e_{4z}^2 - 1) \tag{A8.9}$$

is to be maximized with regard to e_{4y} and e_{4z}.
Partial differentiation of Equation (A8.9) gives

$$P_{e_{4y}} = e_{3z}(1 - e_{2x}) + 2me_{4y} = 0 \tag{A8.10}$$
$$P_{e_{4z}} = -e_{3y}(1 - e_{2x}) + e_{2y}(1 - e_{3x}) + 2me_{4z} = 0 \tag{A8.11}$$

and after eliminating m we obtain

$$(e_{3z}e_{4z} + e_{3y}e_{4y})\,(1 - e_{2x}) - e_{2y}e_{4y}(1 - e_{3x}) = 0 \tag{A8.12}$$

which, together with the conditions of Equation (9.60), gives the solutions

$$e_{4y} = \pm\, e_{3z}\left[\frac{(1 - e_{2x})\,(1 + e_{3x})}{2(1 - e_{2x}e_{3x} - e_{2y}e_{3y})}\right]^{1/2} \tag{A8.13}$$

$$e_{4z} = \pm\, \frac{e_{2y}(1 - e_{3x}) - e_{3y}(1 - e_{2x})}{1 - e_{2x}}\left[\frac{(1 - e_{2x})\,(1 + e_{3x})}{2(1 - e_{2x}e_{3x} - e_{2y}e_{3y})}\right]^{1/2} \tag{A8.14}$$

(plus sign in both or minus sign in both). Insertion into Equation (A8.8) gives

$$V_{max} = \frac{1 - e_{3x}}{6}\,[[2(1 - e_{2x})(1 + e_{3x})(1 - e_{2x}e_{3x} - e_{2y}e_{3y})]^{1/2}$$
$$+ |e_{2y}e_{3z}|] \tag{A8.15}$$

Based on the above, the following is a suitable procedure for satellite selection:

1. First select the satellite with the largest elevation angle.

2. Select the satellite that meets the requirement for an angle of 109.5° most closely from the first one.

3. Select the satellite which, together with numbers 1 and 2, maximizes Equation (A8.15).

4. Select the satellite which together with numbers 1, 2 and 3 maximizes Equation (9.59).

This method of satellite selection is obviously not optimum since not all combinations are tested. The advantage is that it is fast and requires minimal computer capacity in relation to the general method. If four out of n visible satellites are to be selected, the general case gives $n!/4!(n-4)!$ possible combinations, all of which have to be examined. For example, the general case with $n = 6$ gives fifteen combinations, and $n = 9$ gives 126 combinations. The above method, on the other hand, gives only twelve and forty-two combinations respectively. The advantage grows with the number of satellites present. (In addition, the calculations of V_{max} in Equation (A8.15) and $|G_u|$ in Equation (9.59) are much faster than the calculation of GDOP in Equation (9.62).)

Simulations of a large number of constellations have shown (Kihara and Okada, 1984) that the simplified method gives almost as good results as the general one. On an average, GDOP according to the simplified method is only 0.118 larger than the optimum one if the selection includes satellites with elevations >5° only. The probability of higher GDOP values increases, however. The reason for this behaviour is that GDOP (and PDOP) in general varies very slowly as a function of the satellite constellation movements.

It should be added that in some receivers the only requirement with regard to PDOP is that it should not exceed a certain limit. If this limit is not too strict, it may be unnecessary to undertake some kind of optimized satellite selection, as the requirement is often met after trying only a few combinations of the available satellites.

Appendix 9
Basic Kalman filter equations

The following basic equations for Kalman filter solutions in navigation systems, or combinations of navigation systems, are given without derivations. Derivations and further explanations can be found in, for example, Gelb, 1964; Tysso 1986.

The optimum combination of measured and estimated results at time t_k is given by

$$\hat{\mathbf{x}}_k = \bar{\mathbf{x}}_k + \mathbf{K}_k(\mathbf{y}_k - \mathbf{D}_k\bar{\mathbf{x}}_k) \tag{A9.1}$$

where $k = 0, 1, 2, \ldots$

$$\bar{\mathbf{x}}_{k+1} = \boldsymbol{\Phi}_k\hat{\mathbf{x}}_k + \boldsymbol{\Omega}_k\bar{\boldsymbol{\upsilon}}, \ \hat{\mathbf{x}}_0 \text{ given} \tag{A9.2}$$

$$\mathbf{K}_k = \hat{\mathbf{X}}_k\mathbf{D}_k^{\mathrm{T}}\mathbf{W}_k^{-1}, \text{ i.e. the Kalman filter gain} \tag{A9.3}$$

$$\hat{\mathbf{X}}_k = \bar{\mathbf{X}}_k - \bar{\mathbf{X}}_k\mathbf{D}_k^{\mathrm{T}}[\mathbf{D}_k\bar{\mathbf{X}}_k\mathbf{D}_k^{\mathrm{T}} + \mathbf{W}_k]^{-1}\mathbf{D}_k\bar{\mathbf{X}}_k \tag{A9.4}$$

$$\bar{\mathbf{X}}_{k+1} = \boldsymbol{\Phi}_k\hat{\mathbf{X}}_k\boldsymbol{\Phi}_k^{\mathrm{T}} + \boldsymbol{\Omega}_k\mathbf{V}_k\boldsymbol{\Omega}_k^{\mathrm{T}} \tag{A9.5}$$

$\boldsymbol{\Phi}_k$ is referred to as the transition matrix of the state vector $\mathbf{x}_k$, $\boldsymbol{\Omega}_k$ is the transition vector of the noise vector $\mathbf{v}_k$, $\mathbf{y}_k$ is the measurement vector at time t_k, related to the state vector $\mathbf{x}_k$ through the equation

$$\mathbf{y}_k = \mathbf{D}_k\mathbf{x}_k + \mathbf{w}_k \tag{A9.6}$$

where $\mathbf{w}_k$ is the noise vector of the measurement. The covariance matrix of $\mathbf{w}_k$ is

$$\mathbf{W}_k = E[\mathbf{w}_k\mathbf{w}_k^{\mathrm{T}}] \tag{A9.7}$$

and the covariance matrix of the noise vector $\mathbf{v}_k$ is

$$\mathbf{V}_k = E[\mathbf{v}_k \mathbf{v}_k^{\mathsf{T}}] \tag{A9.8}$$

$\bar{\mathbf{X}}_k$ is the covariance matrix of the estimated vector $\mathbf{x}_k$, and $\hat{\mathbf{X}}_k$ is the covariance matrix of the optimized state vector $\hat{\mathbf{x}}_k$. The linearized time-dependence of $\mathbf{\Phi}_k$ and $\mathbf{\Omega}_k$ is

$$\mathbf{\Phi}_k \approx (\mathbf{I} + \Delta t \mathbf{A}_k) \tag{A9.9}$$

and

$$\mathbf{\Omega}_k \approx \Delta t \mathbf{C}_k \tag{A9.10}$$

where

$$\mathbf{A}_k = \frac{\mathrm{d}\dot{\mathbf{x}}_k}{\mathrm{d}\mathbf{x}_k^{\mathsf{T}}} \tag{A9.11}$$

and

$$\mathbf{C}_k = \frac{\mathrm{d}\dot{\mathbf{x}}_k}{\mathrm{d}\mathbf{v}_k^{\mathsf{T}}} \tag{A9.12}$$

i.e.

$$\dot{\mathbf{x}}_k = \mathbf{A}_k \mathbf{x}_k + \mathbf{C}_k \mathbf{v}_k \tag{A9.13}$$

and

$$\dot{\mathbf{y}} = \mathbf{D}_k \dot{\mathbf{x}}_k + \dot{\mathbf{w}}_k \tag{A9.14}$$

The Kalman filter gain can also be written

$$\mathbf{K}_k = \bar{\mathbf{X}}_k \mathbf{D}_k^{\mathsf{T}} (\mathbf{D}_k \bar{\mathbf{X}}_k \mathbf{D}_k^{\mathsf{T}} + \mathbf{W}_k)^{-1} \tag{A9.15}$$

and the optimum covariance matrix can be expressed as

$$\hat{\mathbf{X}}_k = (\mathbf{I} - \mathbf{K}_k \mathbf{D}_k) \bar{\mathbf{X}}_k (\mathbf{I} - \mathbf{K}_k \mathbf{D}_k)^{\mathsf{T}} + \mathbf{K}_k \mathbf{W}_k \mathbf{K}_k^{\mathsf{T}} \tag{A9.16}$$

References

AGARD (1989) 'Kalman filter integration of modern guidance and navigation systems'. *AGARD Lecture Series No. 166.*

Allison, M. T. and Daly, P. (1984) 'Experiences with the Transit navigation satellites at VHF frequencies'. *Space Communication and Broadcasting*, No. 2.

Anodina, T. G. (1982) *International Frequency Registration Board, Circular 1522*, AR 11/A/3, ITU, Genève.

Anodina, T. G. (1988a) 'The Glonass system technical characteristics and performance'. *ICAO FANS Working Paper No. 75*, May.

Anodina, T. G. (1988b) 'Glonass satellite navigation system'. *IMO Sub-Committee on Safety of Navigation. NAV 35/INF.3*, September.

Baltzersen, Ø., Bergh, K., Hals, T. and Forssell, B. (1988) 'Study of NAVSAT for precise positioning'. *ESA Contract Report*, IKU, Trondheim, July.

Bartholomew, C. A. (1978) 'Satellite frequency standards'. *Navigation (USA)*, Vol. 25, No. 2.

Beattie, J. H. (1988) ' The European radionavigation mix in the year 2000'. *NAV'88, Conference of the RIN*, London, March 1988.

Beier, W. and Wolf, M. (1987) *Signalverarbeitung eines digitalen C/A-Code GPS-Empfängers (Signal Processing in a Digital C/A-code GPS Receiver).* SEL, Stuttgart.

Berlin, P. (1988) *Geostationary Application Satellites*. Cambridge University Press.

Beukers, J. M. (1974) 'A review and applications of VLF and LF transmissions for navigation and trading'. *Navigation (USA)*, Vol. 21, No. 2.

Bigelow, S. C. (1983) *A Loran-C Vessel Location System for Traffic Control in the Suez Canal*. Megapulse, Inc., Bedford, Mass.

Blackwell, E. G. (1985) 'Overview of differential GPS methods'. *Navigation (USA)*, Vol. 32, No. 2.

Blanchard, W. F. (1983) 'The continuing development of Transit'. *IEE Conf. Publ.*

Blankenburgh, J. C. and Coldevin, A. J. (1976) 'Results of Doppler satellite test measurements in Norway'. *IKU Publication No. 88*, Trondheim.

Braisted, P. *et al.* (1986) 'Combining Loran and GPS – the best of both worlds. *Navigation (USA)*, Vol. 33, No. 1.

Breien, T. (1979) 'Multipath analysis of ILS glidepath'. *AGARD Conf. Proceedings No. 269.*

Bruckner, D. and Westling, G. (1985) 'Differential Loran-C. Estimator improvement and local system implementation'. *WGA 14th Annual Techn. Symp.*, Santa Barbara, Ca., October.

Copps, E. M. (1984) 'An aspect of the role of the clock in a GPS receiver'. *Navigation (USA)*, Vol. 31, No. 3.

Corrigan, A. L. and Skrivanck, R. A. (1974) *Chart 'Aerospace Environment'.* Air Force Cambridge Research Labs, USAF 1974.

Cox, D. B. (1978) 'Integration of GPS with inertial navigation systems'. *Navigation (USA)*, Vol. 25, No. 2.

Cramér, H. (1946) *Mathematical Methods of Statistics*, Chs. 21–24. Princeton University Press.

Dale, S. A. and Daly, P. (1987) 'An early comparison of the preparational GPS and Glonass satellite navigation systems'. *MELECON '87*, Rome, March.

Dale, S. A. and Daly, P. (1988) 'Development in interpretation of the Glonass navigation satellite data structure'. *IEEE PLANS '88*, Orlando, Fl., November.

Daly, P. (1984) 'Cosmos revisited: the USSR VHF satellite navigation system'. *Space Communication and Broadcasting*, Vol. 2.

Daly, P. (1988) 'Aspects of the Soviet Union's Glonass satellite navigation system'. *The Journal of Navigation*, Vol. 41, No. 2.

Daly, P. and Allison, M. T. (1986) 'Development of a prototype, experimental, single-channel, sequencing Navstar-GPS receiver'. *International Journal of Satellite Communications*, Vol. 4.

Daly, P. and Lennon, G. E. (1989) 'Potential for an integrated GPS/Glonass civil navigation system'. *NAV 89*, London, October.

Daly, P. *et al.* (1988) 'Time transfer with Glonass navigation satellites'. The ION Satellite Div. Techn. Meeting, Colorado Springs, September.

Daly, P. and Perry, G. E. (1986) 'Recent developments with the Soviet Union's VHF satellite navigation system'. *Space Communication and Broadcasting*, No. 4.

Danchik, R. J. (1989) 'Navy Navigation Satellite System status'. *NAV 89*, London, October.

David, P. and Voge, J. (1969) *Propagation of Waves.* Pergamon Press.

Davies, N. G. (1973) 'Performance and synchronisation considerations'. *AGARD LS-58.*

de Bruin, A. C. and van Willigen, D. (1988) 'Loran-C in a hostile environment'. *NAV '88, The 1988 Conf. on the Royal Institute of Navigation*, London, March.

DECCA (1979) *The DECCA Navigator. Principles and Performance of the System.* DECCA Nav. Co., August 1979.

de Lorme, J. F. and Tuppen, A. R. (1975) 'Low-cost airborne Loran-C navigator'. *Electrical Communication*, Vol. 50, No. 4.

Dixon, R. C. (1976) *Spread-spectrum Systems*. John Wiley and Sons.

Doherty, J. T. and Feldman, D. A. (1975) 'Calculator assisted Loran-C controller for time difference error control'. *Institute of Navigation National Marine Meeting*, October.

D' Oliveira, B. (1988) 'Current European navaids'. *NAV '88*, London, March.

Dornheim, A. A. (1988) 'Potential satellite services laboring under conflicting frequency schemes'. *Aviation Week and Space Technology*, 4 January.

Easton, R. L. (1978) 'The navigation technology program'. *Navigation (USA)*, Vol. 25, No. 2.

Eissfeller, B. and Spietz, P. (1989) 'Basic filter concepts for the integration of GPS and an inertial ring-laser–gyro strapdown system'. *Manuscripta Geodaetica*, No. 14.

Elias, A. L. (1985) 'Aircraft approach guidance using relative Loran-C navigation'. *Navigation (USA)*, Vol. 32, No. 1.

Enge, P. *et al.* (1987) 'Coverage of a radio beacon based differential GPS network'. *Navigation (USA)*, Vol. 34, No. 4.

Engen, B. (ed.) (1988) 'Satellittgeodesi i Norge'. ('Satellite geodesy in Norway'). *Working Group Report, Norwegian Mapping Agency*, Hønefoss.

Fanneløp, I. J. (1982) Statiske utladningers innflytelse på navigasjonssystemer i helikoptre (The Influence of Static Discharges on Helicopter Navigation Systems) MSc Thesis, Norwegian Institute of Technology, Division of Telecommunications, Trondheim.

Feess, W. A. and Stephens, S. G. (1987) 'Evaluation of GPS ionospheric time-delay model'. *IEEE Trans.*, Vol. AES-23, No. 3.

Foley, T. M. (1988) 'Space operations begin using Geostar payload'. *Aviation Week and Space Technology*, 25 July.

Forssell, B. (1974) 'Comparative analysis of microwave landing systems with regard to their sensitivity to coherent interference'. AGARD Joint Meeting on Electromagnetic Noise, Interference and Compatibility, Paris, Oct. 1974.

Forssell, B. (1980) 'SARSAT. Beregning av posisjonsfeil ved lokalisering av nødbøyer via satellitt'. ('Computation of position errors in satellite localization of emergency beacons'). *ELAB Report STF44 F80186*, Trondheim, June.

Forssell, B. (1982a) 'Position and position error ellipses for DME ρ–ρ navigation. Aspects on North Sea helicopter routes'. *ELAB Report STF44 A82112*, Trondheim, January.

Forssell, B. (1982b) 'Helicopter operations for Norwegian offshore installations'. *The Journal of Navigation*, Vol. 35, No. 2.

Forssell, B. (1984) 'Navigation in Norwegian offshore activities'. *IEEE PLANS '84*, November.

Forssell, B. (1985) 'Use of differential Loran-C in Norwegian offshore activities'. *The 5th Int. Congr. of the Int. Ass. of the Institutes of Navigation*, Tokyo, October.

Forssell, B. (1986) 'Do we need an international global civil satellite system?' *The 1986 Conf. of The Royal Inst. of Nav.*, Brighton, October.

Forssell, B. (1988) 'Use of satellite signals for high-accuracy positioning'. *The 6th Int. Congr. of the IAIN*, Sydney, Australia, February.

Francisco, S. G. (1984) 'Operational control segment of the GPS'. *IEEE PLANS '84*, San Diego, November.

Frank, R. L. (1960) 'Multiple pulse and phase-code modulation in the Loran-C system'. *IRE Trans. on Aeronautical and Navigational Electronics*, June.

Frank, R. L. (1983) 'Current developments in Loran-C'. *Proc. of the IEEE*, Vol. 71, No. 10, October.

Fried, W. R. and Kayton, M. (eds.) (1969) *Avionics Navigation Systems*, John Wiley and Sons.

FRNP (1988) '1988 Federal Radio Navigation Plan'. *Report No. DOD-4650 and DOT-TSC-RSPA-88-4*.

Gelb, A. (ed.) (1974) *Applied Optimal Estimation*. Analytic Sciences Corp., MIT Press.

Gething, P. J. D. (1978) *Radio Direction-finding and the Resolution of Multicomponent Wave-fields*. Peter Peregrinus Ltd.

Gløersen, C. A. *et al. Navigasjonsteknikk.* (Navigational Techniques.) Teknologisk Forlag, Oslo, 1974.

Goddard, R. B. and Vicksell, F. B. (1986) 'Implementation and performance of the TOT-controlled French Loran-C chain'. *The Wild Goose Association 15th Annual Techn. Symp.*, New Orleans, La., October.

Gold, R. (1967) 'Optimal binary sequences for spread spectrum multiplexing'. *IEEE Trans.*, Vol. IT-13, October.

Green, G. B. *et al.* (1989) 'The GPS 21 primary satellite constellation'. *Navigation (USA)*, Vol. 36, No. 1.

Gupta, R. R. and Morris, P. B. (1986) 'Omega navigation signal characteristics'. *Navigation (USA)*, Vol. 33, No. 3.

Hagle, L. (1988) 'Investigation of the potential application of GPS for precision approaches'. *Navigation (USA)*, Vol. 35, No. 3.

Hansen, E. (1989) Bruk av GPS sammen med treghetsnavigasjon ved kontrollflyging (Integrated Use of GPS and Inertial Navigation for Flight Inspection). MSc Thesis, Norw. Inst. of Technology, Div. of Telecomm., Trondheim.

Harris, R. L. (1973) 'Introduction to spread-spectrum techniques'. *AGARD Lecture Series No. 58*. 1973.

Hatch, R. R. (1978) 'Hyperbolic positioning per se is passé'. IEEE PLANS.

Held, V. and Kricke, K. D. (1985) 'GPS satellite navigation in the urban environment'. *NAV 85*, York, September.

Hernandez, D. (1986) 'Le système Locstar de radiorepérage par satellite' ('The Locstar system for satellite radiodetermination'). *ESA Workshop on Land-Mobile Services by Satellite*, ESTEC, Noordwijk, September.

Hervig, K. (1988) 'Diffstar – a project based on differential GPS in northern Norway'. IEEE PLANS, '86, November.

Heyes, D. D. (1988) 'The Loran-C midcontinent expansion project'. *IEEE*

PLANS '88, Orlando, Florida, November.

Holsen, J. (1986) *Rotasjonsellipsoide og kule. Beregninger og avbildninger (Rotational Ellipsoid and Sphere. Calculations and Mappings)*. The Institute of Geodesy and Photogrammetry, NTH, Trondheim.

Holte, S. (1989) GPS mottakertyper (*GPS Receiver Types*). Kongsberg Navigation, Kongsberg, December.

Hoogenraad, B. (1988) Loran-C Pulse Position Modulation and Demodulation. MSc Thesis, Norwegian Institute of Technology, Trondheim, August.

Hoskins, G. W. and Danchik, R. J. (1984) 'Navy Navigation Satellite System. Status and future'. *NAV '84, The 1984 Conf. of The Royal Institute of Navigation*, London, May.

Howell, J. M. (1987) 'Phased arrays for microwave landing systems'. *Microwave Journal*, Vol. 30, No. 1.

ICAO (1985) *International Standards and Recommended Practices*. Aeronautical Telecommunications. Annex 10. Vol. 1, ICAO.

ICAO (1978) *Report of the All Weather Operations Divisional Meeting*. ICAO DOC 9242, AW078, April 1978.

IMCO (1980) 'Performance standards for differential OMEGA correction transmitting stations'. *IMCO Resolution A.425(XI)*, January 1980.

Johannessen, R. (1987–8) 'International future navigation needs: Options and concerns'. *Navigation (USA)*, Vol. 34, No. 4.

Johannessen, R. (1989) 'Report on the study of a world-wide radionavigation system'. *NAV 35/14, Annex 9, The International Maritime Organization*, January.

Jones, W. E. (1978) 'Envelope-to-cycle difference'. *7th Annual Techn. Symp., Wild Goose Ass.*, New Orleans, October.

Jorgensen, P. S. (1989) 'An assessment of ionospheric effects on the GPS user', *Navigation (USA)*, Vol. 36, No. 2.

Joyce, H. *et al.* (1989) 'Navigation satellite system and payload definition study'. *ESA Contract (8004/88/F/RD) Final Report*, Marconi Space Systems, Racal Research and Italspazio, December.

Kai Fong Lee (1984) *Principles of Antenna Theory*. John Wiley and Sons.

Kalafus, R. (1989) 'GPS integrity channel RTCA working group recommendations'. *Navigation (USA)*, Vol. 36, No. 1.

Kalafus, R. *et al.* (1986) 'Special Committee 104 Recommendations for Differential GPS Service'. *Navigation (USA)*, Vol. 33, No. 1.

Kelly, R. J. (1984) 'System considerations for the new DME/P international standard'. *IEEE Trans.*, Vol. AES-20, No. 1.

Kelso, J. M. (1964) *Radio Ray Propagation in the Ionosphere*. McGraw-Hill.

Kerr, D. E. (1951) *Propagation of Short Radio Waves. MIT Rad. Lab. Series No. 13*, McGraw-Hill.

Kihara, M. and Okada, T. (1984) 'A satellite selection method and accuracy for the Global Positioning System'. *Navigation (USA)*, Vol. 31, No. 1.

Kilgus, C. C. (1969) 'Resonant quadrifilar helix'. *IEEE Trans.*, Vol. AP-17, No. 3.

Klein, D. and Parkinson, B. W. (1984–5) 'The use of pseudo-satellites for improving GPS performance'. *Navigation (USA)*, Vol. 31, No. 4.

Klobuchar, J. A. (1987) 'Ionospheric time-delay algorithm for single-frequency GPS users'. *IEEE Trans.*, Vol. AES-23, No. 3.

Knable, N. and Kalafus, R. M. (1984–5) 'Clock coasting and altimeter error analysis for GPS'. *Navigation (USA)*, Vol. 31, No. 4.

Lamiraux, C. (1988) 'Nouveau récepteur Loran-C à couverture européenne' ('A new Loran-C receiver with European coverage'). *Navigation (France)*, No. 143, July.

Larkin, T. (1988) 'GPS services available to civilian users'. *Proceedings of the ION Satellite Div. Int. Techn. Meeting*, Colorado Springs, September.

Last, J. D. and Linsdall, D. K. (1984) 'Automatic microprocessor-based receivers for the Decca Navigator system'. *The Radio and Electronic Engineer*, Vol. 54, No. 7/8.

Laube, S. J. P. (1987) *GPS Antenna Designs*. Ohio University, Athens, Ohio, May.

Laurent, P. and Nard, G. (1975) 'Le Syledis, Système de radio-localisation précise à courte et moyenne distance' ('Syledis. An accurate radio-positioning system for short and medium ranges'). *Navigation (France)*, Vol. 23, No. 89.

Lennon, G. R. (1989) 'The USSR's Glonass P-code – determination and initial results'. *The 2nd Int. Meeting of the Satellite Div. of The Institute of Navigation*, Colorado Springs, September.

Lopez, A. R. (1982) 'Scanning-beam MLS. Multipath errors and antenna design philosophy'. *IEEE Trans.*, Vol. AP-25, No. 3.

MacKenzie, F. D. (1984) 'A survey of state-of-the-art Loran-C receivers'. *Report No. DDT-CG-N-1-84, Transportation Systems Center*, Cambridge, Mass., June 1984.

Maher, R. A. (1984) 'A comparison of multichannel, multiplex and sequential GPS receivers for air navigation'. *Navigation (USA)*, Vol. 31, No. 2.

Mattos, P. G. (1988) 'A low-cost hand-held GPS navigation system receiver'. *4th Int. Conf. on Sat. Syst. for Mobile Comm. and Nav. IEE Conf. Publ. No. 294*, London, October.

Mattos, P. G. (1989) 'The transputer in satellite communications signal processing. *Applications of Transputers, Satellite Communications*, November.

McGann, E. (1987) 'On GPS expansion'. *The Goose Gazette*, Fall.

Milliken, R. J. and Zoller, C. J. (1978) 'Principle of operation of NAVSTAR and system characteristics'. *Navigation (USA)*, Vol. 25, No. 2.

Moskvin, G. I. and Sorochinsky, V. A. (1989) 'The Glonass satellite navigation system. Navigational aspects'. *NAV 89*, London, October.

Næss, A. (1952) *Navigasjonsformler med sfærisk trigonometri (Navigational Formulae with Spherical Trigonometry)*. Grøndahl and Søn, Oslo.

Napier, M. (1989) 'The integration of satellite and inertial positioning systems'. *The NAV 89 Conf. of the Royal Inst. of Nav.*, London, September.

Nard, G. (1979) 'Les appareils et la navigation Oméga différentiel des années 1980'. ('Differential Omega equipment and navigation in the 1980's'). *Colloque du S.H.O.M. sur la localisation en mer*, Brest, Oct. 1979.

Nard, G. (1984) 'Spread-spectrum concept applied in new accurate medium/ long-range radiopositioning system'. *Sercel Company Report*.

Nard, G. (1989) 'Méthodes de prospection pour la mise en æuvre de stations Oméga différentiel' ('Investigation methods for implementation of differential Omega stations'). *Navigation (France)*, Vol. 32, No. 127.

Nard, G. *et al.* (1989) 'Introduction to the GPS and to Sercel GPS receivers'. *Sercel Company Report*, March.

NATO (1987) *Navstar Global Positioning System (GPS) System Characteristics. Standardization Agreement 4294*, NATO.

Norton, K. A. (1936–7) 'The propagation of radio waves over the surface of the earth and in the upper atmosphere'. *IRE Proc.*, Vol. 24, October 1936, and Vol. 25, September 1937.

Ott, L. and Blanchard, W. (1988) 'The STARFIX satellite navigation system'. *The 4th Int. Conf. on Sat. Syst. for Mobile Comm. and Nav.*, London, October (*IEEE Conf. Publ. No. 294*).

Ould, P. C. and van Wechel, R. J. (1981) 'All-digital GPS receiver mechanization'. *Navigation (USA)*, Vol. 28, No. 3.

Ould, P. C. and van Wechel, R. J. (1986) 'The modular digital approach to GPS receiver design'. *IEEE PLANS '86*, November.

Parkinson, B. W. and Gilbert, S. W. (1983) 'NAVSTAR/GPS: Ten years later'. *IEEE Proc.*, Vol. 71, No. 10.

Perry, G. E. and Wood, C. D. (1976) 'Identification of a navigation satellite system within the Cosmos programme'. *J. of the British Interplanetary Society*, Vol. 29.

Pierce, J. A. (1965) *Omega. IEEE Trans. on Aerospace and Electronic Systems*. Vol. AES-1, No. 6, December.

Pierce, J. A. (1989) 'Invention of Omega'. *Navigation (USA)*, Vol. 36, No. 2.

Piscane, V. L. *et al.* (1973) 'Recent improvements in the Navy Navigation Satellite System'. *Navigation (USA)*, Vol. 20, No. 3.

Poppe Jr, M. C. (1982) 'The Loran-C receiver. A functional description'. *Navigation (USA)*, Vol. 29, No. 1, Spring.

Powell, C. (1982) 'Performance of the Decca Navigator on land'. *IEE Proc.*, Vol. 129, Pt. F, No. 4.

Redlien, H. W. and Kelly, R. J. (1984) 'Microwave landing system: the new international standard'. *Advances in Electronics and Electron Physics*, Vol. 57, Academic Press.

Remondi, B. W. (1986) 'Performing cm-accuracy kinematic surveys in seconds, using GPS. Modelling and results'. *4th Int. Symp. on Satellite Positioning*, Austin, Texas, May.

Reynolds, F. L. (1953) 'An examination of some site and path errors of the Decca Navigator system when used over land'. *Proc. IEE*, 1953, 100, Part III.

Rohde & Schwarz (1980) 'Precision VHF direction finder PA001 for position finding of ships'. *News from Rohde & Schwarz*, Vol. 20, No. 90, 1980/III.

Rosetti, C. (1986) 'NAVSAT in the year 2000'. *The 1986 Conf. of the Royal Inst. of Navigation*, Brighton, October.

Rosetti, C. (1989) 'A civil complement to GPS and Glonass'. *NAV 89, The*

1989 Conf. of the Royal Inst. of Nav., London, October.

Rothblatt, M. A. and van Till, S. W. (1986) 'Radiodetermination satellite service and the process of international regulatory recognition'. *Space Communication and Broadcasting*, Vol. 4.

RTCA (1970) 'A new guidance system for approach and landing'. *SC-117 RTCA, DO 148*. December.

Sæther, K. (1988) Erfaringer med Loran-C i forbindelse med posisjonering offshore (*Experiences with Loran-C in Offshore Positioning*). Offshore Northern Seas, Stavanger, August.

Sæther, K. and Vestmo, E. (1985) 'Loran-C signal stability study in North-Norway'. *WGA 14th Annual Techn. Symp.*, Santa Barbara, California, October.

Samaddar, S. N. (1979) 'The theory of Loran-C ground wave propagation – a review'. *Navigation (USA)*, Vol. 26, No. 3.

Samaddar, S. N. (1980) 'Weather effects on Loran-C propagation'. *Navigation (USA)*, Vol. 27, No. 1.

Scales, W. C. and Swanson, R. (1984) 'Air and sea rescue via satellite systems'. *IEEE Spectrum*.

Schmidt, G. T. *et al.* (1989) 'Kalman filter integration of modern guidance and navigation systems'. *AGARD Lecture Series No. 166*, June.

Shapiro, L. D. (1968) 'Loran-C sky-wave delay measurement'. *IEEE Trans.*, Vol. IM-17, No. 4, December.

Sherman, H. T. and Johnsen, V. L. (1976–7) 'The Loran-C ground station'. *Navigation (USA)*, Vol. 23, No. 4, Winter 1976–77.

Sherod, C. (1989) 'GPS positioned to change our lives and boost our industry'. *Microwave Systems News*, Vol. 19, No. 7.

Sigmond, M. E. (1990) Ionosfæreforstyrrelser ved bruk av GPS (Ionospheric Disturbances of GPS Use). MSc Thesis, Norw. Inst. of Technology.

Smith, M. S. *et al.* (1989) 'Low profile antennas and arrays for satellite navigation'. *Int. Conf. on Antennas and Propagation, Coventry 1989*. IEEE Conference Publication 301.

Snively, L. O. and Osborne, W. P. (1986) *Analysis of the Geostar Position Determination System*. Geostar, Inc.

Spilker, J. J (1978) 'GPS signal structure and performance characteristics'. *Navigation (USA)*, Vol. 25, No. 2.

Stansell, T. A. (1978) *The Transit Navigation Satellite System*. Magnavox.

Stansell, T. A. (1978) 'The MX 1502 Satellite Surveyor'. *Magnavox Report R-5807A*, February.

Stansell, T. A. (1979) 'The MX 1502 Satellite Surveyor'. *The 2nd Int. Geodetic Symp. on Satellite Doppler Positioning*, Austin, January.

Stein, B. A. (1985) 'Satellite selection criteria during altimeter aiding of GPS'. *Navigation (USA)*, Vol. 32, No. 2.

Stenseth, A. (1988) 'Loran-C in north-west Europe – a status report'. *WGA 17th Ann. Techn. Symp.*, Portland, Oreg., October.

Swanson, E. R. (1977) 'Propagation effects on Omega'. *AGARD Conference Proceedings*, CP-209.

Tachita, R. *et al.* (1989) 'Low-cost multi-channel GPS receiver'. *Proceedings of the Second Int. Techn. Meeting of the Satellite Div. of The Institute of Navigation, ION GPS-89*, Colorado Springs, September.

Taguchi, K. *et al.* (1987) 'Position determination at sea using LF radio waves. The effect of a nearby mountain'. *IEEE J. of Oceanic Eng.*, Vol. OE-12, No. 3.

Thomas, Paul D. (1968) *Mathematical Models for Navigation Systems*. R&D Department, US Naval Oceanographic Office.

Thrane, E. V. and Røed Larsen, T. 'Ionosfærens innvirkning på Loran-C i polare strøk'. ('Ionospheric influence on Loran-C in arctic areas'). *Elektro-Elektroteknisk tidsskrift*, Vol. 88, No. 4, February.

Trimble, C. R. (1988) GPS C/A-code receiver. *US Patent No. 4754465*, June.

Tyssø, A. (1986) *Navigasjonssystemer (Navigation Systems)*. Norwegian Institute of Technology, Division of Telecommunications, 1986.

van der Waal, P. W. and van Willigen, D. (1978) 'Hard limiting and sequential detection applied to Loran-C'. *IEEE Trans.*, Vol. AES-14, No. 4, July.

van Dierendonck, A. J. *et al.* (1978) 'The GPS navigation message'. *Navigation (USA)*, Vol. 25, No. 2.

van Willigen, D. (1985) *Hard Limiting and Sequential Detecting Loran-C Sensor*. Delft University Press.

Viehweg, C. S. *et al.* (1988) 'Differential Loran-C project. A status report'. *WGA 17th Ann. Techn. Symp., Portland, Or.*, October.

Wait, J. R. and Spies, K. P. (1964) 'Characteristics of the earth–ionosphere waveguide for VLF radio waves'. *Technical Note 300, National Bureau of Standards*.

Watt, A. D. (1967) *VLF Radio Engineering*. Pergamon Press.

Wenzel, R. J. and Slagle, D. C. (1983) 'Loran-C signal stability study'. *Report No.CG-D-28-83*, US Coast Guard, August.

Westling, G. (1984) 'Joint Soviet–American Loran operations'. *IEEE AES Magazine*, February.

Woolley, D. (1985) 'Preparing for MLS. The hard work begins'. *Interavia*, No. 1.

Yakos, M. *et al.* (1975) *Space Vehicle Navigation Subsystem and NTS PRN Navigation Assembly/User System Segment and Monitor Station*. Rockwell Int. Corp., Space Div., September.

Young, A. C. (1980) Aircraft Navigation Aids – Reducing the Influence of the Ground on the Performance of the ILS Glidepath and VHF Omnirange. Ph.D. Thesis, School of Electrical Engineering, University of Sydney.

Zeltser, M. J. and El-Arini, M. B. (1985) 'The impact of cross-rate interference on Loran-C receivers'. *IEEE Trans.*, Vol. AES 21, No. 7, January.

Index